SpringerBriefs on PDEs and Data Science

Editor-in-Chief

Enrique Zuazua, Department of Mathematics, University of Erlangen-Nuremberg, Erlangen, Germany

Series Editors

Irene Fonseca, Department of Mathematical Sciences, Carnegie Mellon University, Pittsburgh, USA

Franca Hoffmann, Hausdorff Center for Mathematics, University of Bonn, Bonn, Germany

Shi Jin, Institute of Natural Sciences, Shanghai Jiao Tong University, Shanghai, China

Juan J. Manfredi, Department of Mathematics, University Pittsburgh, Pittsburgh, USA

Emmanuel Trélat, CNRS, Laboratoire Jacques-Louis Lions, Sorbonne University, PARIS CEDEX 05, France

Xu Zhang, School of Mathematics, Sichuan University, Chengdu, China

W0111541

SpringerBriefs on PDEs and Data Science targets contributions that will impact the understanding of partial differential equations (PDEs), and the emerging research of the mathematical treatment of data science.

The series will accept high-quality original research and survey manuscripts covering a broad range of topics including analytical methods for PDEs, numerical and algorithmic developments, control, optimization, calculus of variations, optimal design, data driven modelling, and machine learning. Submissions addressing relevant contemporary applications such as industrial processes, signal and image processing, mathematical biology, materials science, and computer vision will also be considered.

The series is the continuation of a former editorial cooperation with BCAM, which resulted in the publication of 28 titles as listed here: www.springer.com/gp/mathematics/bcam-springerbriefs

Weijun Meng • Jingtao Shi • Jiongmin Yong

Time-Delayed Linear Quadratic Optimal Control Problems

Springer

Weijun Meng
Academy of Mathematics and Systems
Science, Chinese Academy of Sciences
Beijing, Beijing, China

Jingtao Shi
School of Mathematics
Shandong University
Jinan, Shandong, China

Jiongmin Yong
Department of Mathematics
University of Central Florida
ORLANDO, FL, USA

ISSN 2731-7595 ISSN 2731-7609 (electronic)
SpringerBriefs on PDEs and Data Science
ISBN 978-981-96-1896-5 ISBN 978-981-96-1897-2 (eBook)
https://doi.org/10.1007/978-981-96-1897-2

© The Editor(s) (if applicable) and The Author(s), under exclusive license to Springer Nature Singapore Pte Ltd. 2025

This work is subject to copyright. All rights are solely and exclusively licensed by the Publisher, whether the whole or part of the material is concerned, specifically the rights of translation, reprinting, reuse of illustrations, recitation, broadcasting, reproduction on microfilms or in any other physical way, and transmission or information storage and retrieval, electronic adaptation, computer software, or by similar or dissimilar methodology now known or hereafter developed.
The use of general descriptive names, registered names, trademarks, service marks, etc. in this publication does not imply, even in the absence of a specific statement, that such names are exempt from the relevant protective laws and regulations and therefore free for general use.
The publisher, the authors and the editors are safe to assume that the advice and information in this book are believed to be true and accurate at the date of publication. Neither the publisher nor the authors or the editors give a warranty, expressed or implied, with respect to the material contained herein or for any errors or omissions that may have been made. The publisher remains neutral with regard to jurisdictional claims in published maps and institutional affiliations.

This Springer imprint is published by the registered company Springer Nature Singapore Pte Ltd.
The registered company address is: 152 Beach Road, #21-01/04 Gateway East, Singapore 189721, Singapore

If disposing of this product, please recycle the paper.

To Our Parents
Haiquan Meng and Runai Yang
Yuekun Shi and Xiuzhi Qu
Wenyao Yong and Xiangxia Chen

Preface

Classical linear-quadratic (LQ, for short) optimal control theory was born at the end of 1950s and by now it has been well-understood. In the recent years, the open-loop and closed-loop solvability has been investigated thoroughly, for both deterministic and stochastic problems, which makes the standard LQ theory more close to perfect. On the other hand, the theory of LQ problems with delays seems to be much pre-mature, mainly because of its infinite dimensional nature. To treat this situation, by now, it is almost standard that people lift the system so that the original problem becomes an LQ optimal control problem in an infinite dimensional space without delays. Then one could try to establish an LQ theory for such a problem. After that, one would try to come back to the original finite dimensional space, in which the results in the infinite dimensional space would have the corresponding representation. Different from the existing literature, in the current book, we are going to present a theory of deterministic LQ problem with delays which has several new features. Let us now elaborate them.

(1) Our system is time-varying, with both the state equation and the cost functional being allowed to include discrete and distributed delays, both in the state and the control. The delays are described by the integrals with respect to some Radon measures on $[-\delta, 0]$. First, we will present some basic properties of the state equation, including the well-posedness, and the variation of constants formula. Then, we will lift the state equation with delays to one in a suitable infinite dimensional Hilbert space, so that the original LQ problem in $\mathbb{R}^n$ with delays becomes one in the Hilbert space without delays. It is worth mentioning that the lifted infinite dimensional control problem contains a new control operator, which makes the lifted problem not a standard infinite-dimensional optimal control problem. Therefore, the existing approaches in the literature do not apply. We are going to take a different approach to overcome the difficulties caused by the new control operator. Based on the above preparations, adopted from Sun–Yong [1], we introduce the notions of open-loop solvability, closed-loop solvability and closed-loop representation of open-loop optimal control for the lifted infinite dimensional control problem.

(2) The open-loop solvability of the lifted problem is characterized by the solvability of a system of forward-backward integral evolution equations (or a two-point boundary value problem of integral evolution equations) and the convexity condition of the cost functional. The proof of which is by no means trivial because the Radon measure and the new control operator in the lifted problem bring difficulties in seeking the adjoint equations. Moreover, we will return to the original finite dimensional problem to reveal its open-loop solvability. It is surprising that the adjoint equations involve some coupled partial differential equations, which is significantly different from that in the literature, where the adjoint equations are all some anticipated backward ordinary differential equations.
(3) The closed-loop solvability is characterized by the solvability of three equivalent integral operator-valued Riccati equations and two equivalent backward integral evolution equations which are easier to handle than the differential operator-valued Riccati equations used in the literature to study similar problems. We will successfully go back to the original finite dimensional problem with delays and obtain the closed-loop solvability of the problem through some coupled matrix-valued Riccati equations as well as coupled partial differential equations.
(4) The closed-loop representation of open-loop optimal control is presented through three equivalent integral operator-valued Riccati equations. Since the new control operator makes the lifted problem not a standard infinite-dimensional optimal control problem, we will use certain approximation arguments. Finally, we will again return to the original finite dimensional problem with delays and present the representation via certain coupled matrix-valued Riccati equations.

For readers' convenience, we have made some efforts to try to write the book self-contained. We provide considerably detailed proofs for most of the results. It is expected that the book will be useful for scholars who are interested in the topic.

Beijing, China — Weijun Meng
Jinan, China — Jingtao Shi
Orlando, FL, USA — Jiongmin Yong
May 2024

Frequently Used Notations

1. $\mathbb{R}^{n\times m}$: Euclidean space of all $n \times m$ real matrices; $\mathbb{R}^n := \mathbb{R}^{n\times 1}$.
2. E: Banach space, could be a Hilbert space, or even a Euclidean space.
3. $\|\cdot\|_E$: the norm in E. Specially, if E is the Euclidean space, write $\|\cdot\|_E$ simply as $|\cdot|$.
4. $\langle \cdot, \cdot \rangle_E$: the inner product of E if it is a Hilbert space. Specially, if E is the Euclidean space, write $\langle \cdot, \cdot \rangle_E$ simply as $\langle \cdot, \cdot \rangle$.
5. E^*: the dual space of Banach space E.
6. $\langle \cdot, \cdot \rangle_{E^*,E}$: the duality pairing between E^* and E, when E is a Banach space.
7. $A^\top$: the transpose of vectors or matrices A.
8. $\mathbb{S}^n$: the space of all $n \times n$ symmetric matrices.
9. I: the identity map/matrix with appropriate domain/dimension.
10. $\mathscr{L}(H, U)$: the real Banach space of all continuous linear maps; $\mathscr{L}(H) := \mathscr{L}(H, H)$.
11. A^*: the adjoint operator of $A \in \mathscr{L}(H, U)$.
12. T: the finite time horizon.
13. $\Delta_*[0, T] := \{(t, s) \in [0, T]^2 \,|\, 0 \leqslant s \leqslant t \leqslant T\}$, the lower triangle domain.
14. $L^p(F; E) := \Big\{\phi : F \to E \;\Big|\; \phi \text{ is measurable}, \displaystyle\int_F \|\phi(t)\|_E^p dt < \infty\Big\}$, $p \in [1, \infty)$.
15. $L^\infty(F; E) := \Big\{\phi : F \to E \;\Big|\; \phi \text{ is measurable}, \displaystyle\sup_{t\in F} \|\phi(t)\|_E < \infty\Big\}$.
16. $C(F; E) := \big\{\phi(\cdot) \in L^\infty(F; E) \;\big|\; \phi(\cdot) \text{ is continuous}\big\}$.
17. $H^1(F; E) := \big\{\phi(\cdot) \in L^2(F; E) \;\big|\; \phi_t(\cdot) \in L^2(F; E),$ where $\phi_t(\cdot)$ indicates the distributional derivative$\big\}$.
18. $L^p(a, b; E) := L^p([a, b]; E)$.
19. $L^\infty(a, b; E) := L^\infty([a, b]; E)$.
20. $H^1(a, b; E) := H^1([a, b]; E)$.
21. $\|\cdot\|_{L^2}$: the norm in $L^2(F; E)$.
22. $\langle \cdot, \cdot \rangle_{L^2}$: the inner product in $L^2(F; E)$.
23. $\|\cdot\|_{H^1}$: the norm in the Sobolev space $H^1(F; E)$.

24. $\langle \cdot, \cdot \rangle_{H^1}$: the inner product in the Sobolev space $H^1(F; E)$.
25. $L^2 := L^2(-\delta, 0; \mathbb{R}^k)$, $k = n, m$.
26. $M^2 := \mathbb{R}^n \times L^2$, $Z := M^2 \times L^2$.
27. $[p]^0$: the component in $\mathbb{R}^n$ of $p \in M^2$.
28. $[p]^1$: the component in L^2 of $p \in M^2$.

Reference

1. Sun, J.R., Yong, J.M.: Stochastic Linear-Quadratic Optimal Control Theory: Open-Loop and Closed-Loop Solutions. Springer, Berlin (2020)

Contents

Chapter 1
Introduction

In this chapter, we formulate the control problems with their motivations that we will investigate. Some preliminary results will be presented followed by some historic comments.

1.1 Motivations

Let us begin with the following controlled linear *ordinary differential equation* (ODE, for short):

$$\begin{cases} \dot{X}(t) = A(t)X(t) + B(t)u(t), & t \in [0, T], \\ X(0) = x. \end{cases} \tag{1.1.1}$$

In the above, $X(\cdot)$ is called the *state trajectory* valued in an n-dimensional Euclidean space $\mathbb{R}^n$ with the *initial state* x, and $u(\cdot)$ is called the *control function* valued in another Euclidean space $\mathbb{R}^m$, $A(\cdot)$ and $B(\cdot)$ are some suitable matrix-valued maps, called the *coefficients*. In application, components of $X(\cdot)$ could be the quantity of various objects, which follow some dynamics described by the *state equation* (1.1.1), under the control action $u(\cdot)$. One may think these to be certain production of products, inventory of goods, locations of moving objects, to mention a few. Different control $u(\cdot)$ will lead to different *output* $X(\cdot)$, and there should be many controls under which the goal of the output could be met, then one would like to

© The Author(s), under exclusive license to Springer Nature Singapore Pte Ltd. 2025

W. Meng et al., *Time-Delayed Linear Quadratic Optimal Control Problems*, SpringerBriefs on PDEs and Data Science,
https://doi.org/10.1007/978-981-96-1897-2_1

choose the best control under suitable sense. More precisely, people would like to introduce some criterion to measure the performance of the control. The following is a popular one:

$$
\begin{aligned}
J(x;u(\cdot)) = \int_0^T & \big[\langle Q(t)X(t), X(t)\rangle + \langle R(t)u(t), u(t)\rangle\big]dt \\
& + \langle GX(T), X(T)\rangle,
\end{aligned}
\tag{1.1.2}
$$

where the *weighting coefficients* $Q(\cdot)$, $R(\cdot)$ are symmetric matrix-valued functions and G is a symmetric matrix. On the right-hand side of the above, the integral term is called the *running cost* and the last term is called the *terminal cost*. The classical LQ optimal control problem can be stated as follows.

Problem (LQ) For any given $x \in \mathbb{R}^n$, find a $u^*(\cdot) \in \mathscr{U}[0,T]$ such that

$$
J(x;u^*(\cdot)) = \inf_{u(\cdot)\in\mathscr{U}[0,T]} J(x;u(\cdot)) \equiv V(x).
$$

Here, $\mathscr{U}[0,T]$ is the class of all *admissible controls*, and $V(\cdot)$ is called the *value function*. In the above case, $u^*(\cdot)$ is called an *open-loop optimal control*, the corresponding state trajectory, denoted by $X^*(\cdot)$, is called the *open-loop optimal state trajectory*, and $(X^*(\cdot), u^*(\cdot))$ is called an *open-loop optimal pair*.

Let us state the following standard hypothesis.

(H1.1) Let $A(\cdot), B(\cdot), Q(\cdot), R(\cdot)$ be measurable and bounded. Moreover, for almost every $t \in [0,T]$, $Q(t)$, $R(t)$ and G are symmetric such that

$$
G \geqslant 0, \quad Q(t) \geqslant 0, \quad R(t) \geqslant \alpha I > 0, \quad t \in [0,T],
$$

meaning that $R(t)$ is uniformly positive definite and $Q(t), G$ are positive semi-definite.

The following is a standard result.

Proposition 1.1.1 *Let* (H1.1) *hold. Then* Problem (LQ) *admits a unique open-loop optimal control* $u^*(\cdot) \in \mathscr{U}[0,T]$, *which has the following representation:*

$$
u^*(t) = -R(t)^{-1}B(t)^\top P(t)X^*(t), \qquad t \in [0,T],
$$

where $P(\cdot)$ *is the solution to the following differential Riccati equation:*

$$
\begin{cases}
\dot{P}(t) + P(t)A(t) + A(t)^\top P(t) + Q(t) \\
\qquad\qquad -P(t)B(t)R(t)^{-1}B(t)^\top P(t) = 0, \quad t \in [0,T], \\
P(T) = G,
\end{cases}
$$

and $X^*(\cdot)$ is the open-loop optimal trajectory satisfying the following closed-loop system:

$$\begin{cases} \dot{X}^*(t) = \big(A(t) - B(t)R(t)^{-1}B(t)^\top P(t)\big)X^*(t), & t \in [0, T], \\ X^*(0) = x. \end{cases}$$

In this case, the value function $V(\cdot)$ admits the following representation:

$$V(x) = \langle P(0)x, x\rangle, \qquad \forall\, x \in \mathbb{R}^n.$$

The above is a well-known result for Problem (LQ) associated with (1.1.1)–(1.1.2). It is satisfactory. However, Problem (LQ) is very ideal and it might be a little oversimplified, which might not be able to cover many important situations in applications. We now explore them in some extent.

First of all, the situation of the state at the current time, say, t, might be affected by that of the state in the past. This is due to the possible memory existing in the dynamic system. Such kind of memory could be continuous and/or discrete in time. For example, the demand of certain goods has a very clearly seasonal memory, which could be continuous in time (called *continuous delay* or *distributed delay*) and also the demand at, say, weekends, holidays, etc., could be quite different from other days, exhibiting the discrete type memory (called *discrete delay*). Mathematically, this can be described by introducing delays of the state in the dynamics (1.1.1). Here is one possible way of modeling it: ($\delta > 0$)

$$\dot{X}(t) = A(t)X(t) + \int_{-\delta}^{0} A_0(t,\theta)X(t+\theta)d\theta + A_1(t)X(t-\delta) + B(t)u(t).$$

On the other hand, in practice, when the controller has decided to apply a control action, it will take some time to exercise the action. Also, when the controller applies a control action, it will take some time before the action affects the state. Therefore, the delay in the control is not avoidable. If we take such a consideration into account, the state equation might become (with $\delta, \delta' > 0$)

$$\begin{aligned} \dot{X}(t) = A(t)X(t) &+ \int_{-\delta}^{0} A_0(t,\theta)X(t+\theta)d\theta + A_1(t)X(t-\delta) \\ &+ B(t)u(t) + \int_{-\delta'}^{0} B_0(t,\theta)u(t+\theta)d\theta + B_1(t)u(t-\delta'). \end{aligned}$$

Having updated the state equation, the quadratic cost functional should be changed accordingly. This will then lead us to the general formulation of our LQ problem in the next section.

1.2 Problem Formulation

We begin with the following definition.

Definition 1.2.1 Let $(E, \mathscr{B}(E))$ be a measurable space with $E \subseteq \mathbb{R}^d$ being a non-empty Borel measurable set and $\mathscr{B}(E)$ being the Borel σ-field of E.

(i) A (signed) measure μ on $(E, \mathscr{B}(E))$ is called a *(signed) Radon measure* if for any compact set $C \subseteq E$, $|\mu(C)| < \infty$. If $\mu : \mathscr{B}(E) \to \left(\overline{\mathbb{R}}\right)^{n\times m}$ (with $\overline{\mathbb{R}} = [-\infty, +\infty]$) such that each component is a signed Radon measure, then μ is called $\left(\overline{\mathbb{R}}\right)^{n\times m}$*-valued signed Radon measure*.

(ii) A measure μ on $(E, \mathscr{B}(E))$ is said to be *diffuse* or *atomless* if for any $x \in E$, $\mu(\{x\}) = 0$.

(iii) A set $E_0 \in \mathscr{B}(E)$ is called the *support* of μ if E_0 is the smallest Borel measurable set such that for any Borel measurable set $\widetilde{E} \subseteq E \setminus E_0$, $\mu(\widetilde{E}) = 0$.

The following is a well-known decomposition theorem for (signed) Radon measures.

Proposition 1.2.2 *Any (signed) Radon measure μ on measurable space $(E, \mathscr{B}(E))$ can be decomposed into a sum of a diffuse (signed) measure μ_0 and a positive linear combination of Dirac measure, i.e.,*

$$\mu = \mu_0 + \sum_{j\geqslant 1} b_j \mu_j, \qquad b_j > 0, \quad j \geqslant 1.$$

Moreover, the support of each μ_j is contained in that of μ.

Now, we present the formulation of the problem. First, for any given $s \in [0, T)$ and $\delta > 0$, we formally consider the following general controlled linear *ordinary differential delay equation* (ODDE, for short) on $[s, T]$:

$$\begin{cases} \dot{X}(t) = \displaystyle\int_{[-\delta,0]} \big[\widetilde{A}(t, d\theta)X(t+\theta) + \widetilde{B}(t, d\theta)u(t+\theta)\big] + b(t), \\ \qquad\qquad\qquad\qquad\qquad\qquad \text{a.e. } t \in [s, T], \\ X(s) = x, \quad X(t) = \varphi(t-s), \quad t \in [s-\delta, s), \\ u(t) = \psi(t-s), \quad t \in [s-\delta, s), \end{cases} \tag{1.2.1}$$

where $X(\cdot)$ is the state and $u(\cdot)$ is the control, taking values in $\mathbb{R}^n$ and $\mathbb{R}^m$, respectively. For any $t \in [0, T]$, $\widetilde{A}(t\,, \cdot)$ and $\widetilde{B}(t\,, \cdot)$ are some signed Radon measures, defined on $(-\infty, 0]$, supported on $[-\delta, 0]$, taking values in $\left(\overline{\mathbb{R}}\right)^{n\times n}$ and $\left(\overline{\mathbb{R}}\right)^{n\times m}$, respectively. The $\mathbb{R}^n$-valued function $b(\cdot)$ is called a *nonhomogeneous term*, and $x \in \mathbb{R}^n$ is the *initial state*, $\varphi(\cdot) \in L^2(-\delta, 0; \mathbb{R}^n)$ and $\psi(\cdot) \in L^2(-\delta, 0; \mathbb{R}^m)$ are called the *history trajectories* of the state and the control, respectively. The term $\int_{[-\delta,0]} \widetilde{A}(t, d\theta)X(t+\theta)$ can cover the term without delays (by choosing it to have an atom at $\theta = 0$), and all possible discrete and distributed delays of the state.

Similarly, $\int_{[-\delta,0]} \widetilde{B}(t, d\theta)u(t+\theta)$ can cover various cases for the control. Below are possible forms for them:

$$\int_{[-\delta,0]} \widetilde{A}(t, d\theta)X(t+\theta) = A_0(t)X(t) + \sum_{i=1}^{N} A_i(t)X(t+\theta_i) + \int_{-\delta}^{0} A^0(t,\theta)X(t+\theta)d\theta,$$
$$\int_{[-\delta,0]} \widetilde{B}(t, d\theta)u(t+\theta) = B_0(t)u(t) + \sum_{i=1}^{N} B_i(t)u(t+\theta_i) + \int_{-\delta}^{0} B^0(t,\theta)u(t+\theta)d\theta,$$

with

$$-\delta = \theta_N < \theta_{N-1} < \cdots < \theta_1 < \theta_0 = 0.$$

Here $A_i(\cdot)$, $A^0(\cdot\,,\cdot)$, $B_i(\cdot)$, $B^0(\cdot\,,\cdot)$ are matrix-valued functions of appropriate dimensions. For such a case, the state $X(\cdot)$ has discrete delays determined by θ_i and $A_i(\cdot)$, and has the distributed delays determined by $A^0(\cdot\,,\cdot)$. The situation for the control is similar. The state and the control may have discrete delays at different schedules. However, by allowing some of $A_i(\cdot)$ and/or $B_i(\cdot)$ to be zero, we can assume θ_i to exhaust all possible discrete delays of the state and the control.

Next, let $\mu(d\theta)$ be a fixed Radon measure on $(-\infty, 0]$ supported on $[-\delta, 0]$ and suppose that the signed Radon measures $\widetilde{A}(t, d\theta)$ and $\widetilde{B}(t, d\theta)$ are absolutely continuous with respect to $\mu(d\theta)$, for all $t \in [0, T]$. Then, by Radon-Nikodým's theorem, we have

$$\widetilde{A}(t, d\theta) = A(t,\theta)\mu(d\theta), \qquad \widetilde{B}(t, d\theta) = B(t,\theta)\mu(d\theta),$$

for some matrix-valued functions $A(\cdot\,,\cdot)$ and $B(\cdot\,,\cdot)$. With such a reduction, we may rewrite (1.2.1) as follows:

$$\begin{cases} \dot{X}(t) = \displaystyle\int_{[-\delta,0]} \big[A(t,\theta)X(t+\theta) + B(t,\theta)u(t+\theta)\big]\mu(d\theta) + b(t), \\ \qquad\qquad\qquad\qquad\qquad\qquad\qquad\qquad \text{a.e. } t \in [s, T], \\ X(s) = x, \quad X(t) = \varphi(t-s), \quad t \in [s-\delta, s), \\ u(t) = \psi(t-s), \quad t \in [s-\delta, s), \end{cases} \tag{1.2.2}$$

for some matrix-valued functions $A(\cdot\,,\cdot)$, $B(\cdot\,,\cdot)$, and $\mu(\cdot)$ is a scalar-valued (nonnegative) Radon measure with the support being $[-\delta, 0]$. The state equation above is understood as the following integral equation:

$$X(t) = x + \int_s^t \Big(\int_{[-\delta,0]} \big[A(r,\theta)X(r+\theta) + B(r,\theta)u(r+\theta)\big]\mu(d\theta) + b(r)\Big)dr.$$

Note that it is reasonable to assume the continuity of $X(\cdot)$ in $[s-\delta, T]$ (posing some compatible conditions on the initial state x and its history function $\varphi(\cdot)$), together with the continuity of $A(\cdot\,,\cdot)$. Thus, the integral of $A(r,\theta)X(r+\theta)$ with respect to $\mu(d\theta)dr$ is well-defined. However, the control $u(\cdot)$ is usually assumed to be merely square integrable (with respect to the Lebesgue measure). Therefore, the integral of $B(r,\theta)u(r+\theta)$ with respect to $\mu(d\theta)dr$ needs to be made precise. To this end, let us first assume that $u(\cdot)$ patched with its history function $\psi(\cdot)$, is continuous on $[s-\delta, T]$, and $B(\cdot\,,\cdot)$ is also continuous. Then Fubini's Theorem is applicable. Our idea is to exchange the order of the integration, then make use of the translation invariance of the Lebesgue measure to transform $u(r+\theta)$ to $u(\tau)$. Afterwards, we exchange the order of integration back. More precisely, we have the following (noting μ has a support in $[-\delta, 0]$)

$$
\begin{aligned}
&\int_s^t\int_{[-\delta,0]} B(r,\theta)u(r+\theta)\mu(d\theta)dr = \int_{[-\delta,0]}\int_s^t B(r,\theta)u(r+\theta)dr\,\mu(d\theta)\\
&=\int_{[-\delta,0]}\int_s^{(s-\theta)\wedge t} B(r,\theta)\psi(r+\theta-s)dr\,\mu(d\theta)\\
&\qquad+\int_{[s-t,0]}\int_{(s-\theta)\wedge t}^{t} B(r,\theta)u(r+\theta)dr\,\mu(d\theta)\quad (s-\theta<t \iff \theta>s-t)\\
&=\int_{[-\delta,0]}\int_\theta^{(\theta+t-s)\wedge 0} B(\tau'-\theta+s,\theta)\psi(\tau')d\tau'\mu(d\theta)\qquad (\tau'=r+\theta-s)\\
&\qquad+\int_{[s-t,0]}\int_{(t+\theta)\wedge s}^{t+\theta} B(\tau-\theta,\theta)u(\tau)d\tau\,\mu(d\theta)\qquad (\tau=r+\theta).
\end{aligned}
$$

Consequently, by Fubini's Theorem again,

$$
\begin{aligned}
&\int_s^t\int_{[-\delta,0]} B(r,\theta)u(r+\theta)\mu(d\theta)dr\\
&=\int_{-\delta}^0\Big(\int_{[\tau'-(t-s),\tau']} B(\tau'-\theta+s,\theta)\mu(d\theta)\Big)\psi(\tau')d\tau'\\
&\qquad+\int_s^t\Big(\int_{[\tau-t,0]} B(\tau-\theta,\theta)\mu(d\theta)\Big)u(\tau)d\tau.
\end{aligned}
\tag{1.2.3}
$$

Note that Fig. 1.1 is in the r-θ plan, which is used to exchange the integration order from $\mu(d\theta)dr$ to $dr\mu(d\theta)$. Figure 1.2 is in the τ-θ (τ'-θ) plan, which is used to exchange the integration order from $d\tau\mu(d\theta)$ to $\mu(d\theta)d\tau$ (from $d\tau'\mu(d\theta)$ to $\mu(d\theta)d\tau'$). We see that the right-hand side of the above is well-defined for any

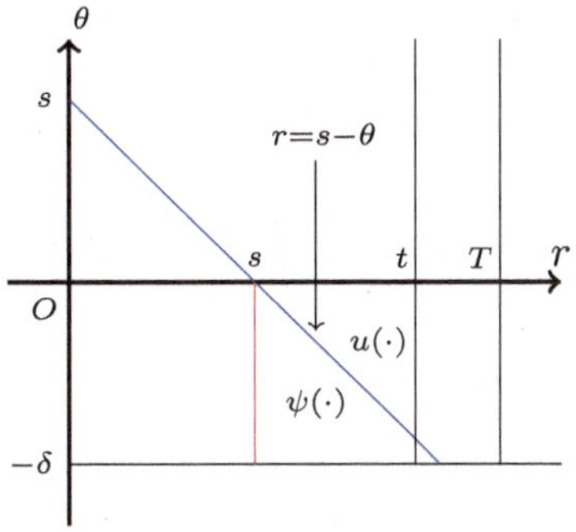

Fig. 1.1 r-θ plan

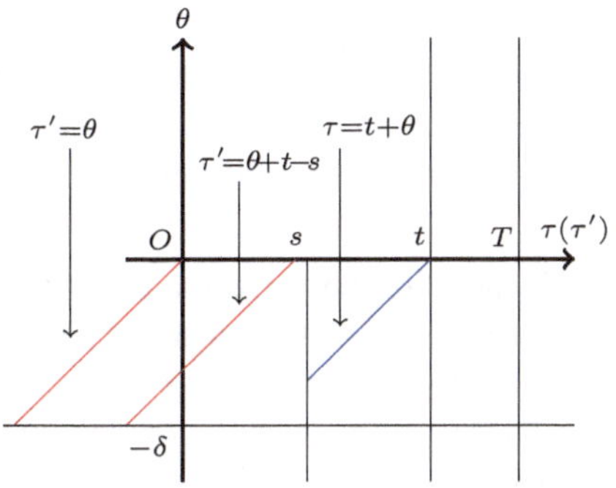

Fig. 1.2 τ-θ (τ'-θ) plan

$\psi(\cdot) \in L^2(-\delta, 0; \mathbb{R}^m)$ and $u(\cdot) \in L^2(s, T; \mathbb{R}^m)$, provided $B(\cdot\,,\cdot)$ is continuous. Likewise, one has

$$
\begin{aligned}
&\int_s^t \int_{[-\delta,0]} A(r,\theta)X(r+\theta)\mu(d\theta)dr \\
&\quad = \int_{-\delta}^0 \Big(\int_{[\tau-(t-s),\tau]} A(\tau-\theta+s,\theta)\mu(d\theta)\Big)\varphi(\tau)d\tau \\
&\qquad + \int_s^t \Big(\int_{[\tau-t,0]} A(\tau-\theta,\theta)\mu(d\theta)\Big)X(\tau)d\tau,
\end{aligned}
\tag{1.2.4}
$$

which is well-defined for all $\varphi(\cdot) \in L^2(-\delta, 0; \mathbb{R}^n)$ and $X(\cdot) \in L^2(s, T; \mathbb{R}^n)$. Then, the state Eq. (1.2.2) can be understood as follows:

$$
\begin{aligned}
X(t) = x &+ \int_{-\delta}^0 \Big[\Big(\int_{[\tau-(t-s),\tau]} A(\tau-\theta+s,\theta)\mu(d\theta)\Big)\varphi(\tau) \\
&+ \Big(\int_{[\tau-(t-s),\tau]} B(\tau-\theta+s,\theta)\mu(d\theta)\Big)\psi(\tau)\Big]d\tau \\
&+ \int_s^t \Big[\Big(\int_{[\tau-t,0]} B(\tau-\theta,\theta)\mu(d\theta)\Big)u(\tau) + b(\tau)\Big]d\tau \\
&+ \int_s^t \Big(\int_{[\tau-t,0]} A(\tau-\theta,\theta)\mu(d\theta)\Big)X(\tau)d\tau.
\end{aligned}
\tag{1.2.5}
$$

In Sect. 2.1, we would like to study the well-posedness of (1.2.5) and furthermore give its explicit solution (variation of constants formula), thus let's express $X(\cdot)$ as a linear integral term of $X(\cdot)$ and a nonhomogeneous term:

$$X(t) = x + \int_s^t \int_{(s-\tau,0]} A(\tau,\theta)X(\tau+\theta)\mu(d\theta)d\tau + \int_s^t f(\tau)d\tau,\ t \in [s,T], \tag{1.2.6}$$

where

$$\begin{aligned} f(\tau) &\equiv f(\tau; s, x, \varphi(\cdot), \psi(\cdot), u(\cdot)) \\ &= \Big[\mu(\{s-\tau\})A(\tau, s-\tau)x + \int_{[-\delta,s-\tau)} A(\tau,\theta)\varphi(\tau+\theta-s)\mu(d\theta) \\ &\quad + \int_{[-\delta,s-\tau)} B(\tau,\theta)\psi(\tau+\theta-s)\mu(d\theta)\Big]\mathbf{1}_{[s,s+\delta]}(\tau) \\ &\quad + \int_{[s-\tau,0]} B(\tau,\theta)u(\tau+\theta)\mu(d\theta)+b(\tau). \end{aligned}$$

To measure the performance of the control $u(\cdot)$, we need to introduce a proper cost functional. Note that the state trajectory $X(\cdot)$ is defined on $[s,T]$. We would like $X(\cdot)$ to have a desired behavior over this interval, and at the same time, from the state equation, at some special moments the discrete delays play important roles and need to be treated differently from other time moments. One should also spend as little energy as possible to achieve these goals for the state. Hence, the following is such a proper cost functional:

$$\begin{aligned} &J(s, x, \varphi(\cdot), \psi(\cdot); u(\cdot)) \\ &= \int_s^T \Big[\int_{[-\delta,0]^2} \Big(\langle Q(t,\theta,\theta')X(t+\theta), X(t+\theta')\rangle \\ &\qquad +2\langle S(t,\theta,\theta')X(t+\theta), u(t+\theta')\rangle \\ &\qquad +\langle R(t,\theta,\theta')u(t+\theta), u(t+\theta')\rangle\Big)\nu(d\theta)\nu(d\theta') \\ &\quad +2\int_{[-\delta,0]} \Big(\langle q(t,\theta), X(t+\theta)\rangle + \langle \rho(t,\theta), u(t+\theta)\rangle\Big)\nu(d\theta)\Big]dt \\ &\quad + \int_{[-\delta,0]^2} \langle G(\theta,\theta')X(T+\theta), X(T+\theta')\rangle\tilde{\nu}(d\theta)\tilde{\nu}(d\theta') \\ &\quad +2\int_{[-\delta,0]} \langle g(\theta), X(T+\theta)\rangle\tilde{\nu}(d\theta), \end{aligned} \tag{1.2.7}$$

where $\nu(\cdot)$, $\tilde{\nu}(\cdot)$ are (scalar-valued) Radon measures. Note that in the above, we allow the linear terms on $X(\cdot)$ and $u(\cdot)$ to appear. Next we want to illustrate that (1.2.7) is well-defined. Suppose all coefficients of (1.2.7) are continuous, by $u(\cdot) \in$

$L^2(s, T; \mathbb{R}^m)$ and $\psi \in L^2(-\delta, 0; \mathbb{R}^m)$, for all $(\theta, \theta') \in [-\delta, 0]^2$, we have (without loss of generality, assuming $\theta \leqslant \theta'$)

$$\begin{aligned}
&\int_{[-\delta,0]^2} \int_s^T \big|\big\langle R(t,\theta,\theta')u(t+\theta), u(t+\theta')\big\rangle\big| dt\nu(d\theta)\nu(d\theta') \\
&\leqslant K\Big(\int_{[-\delta,0]^2} \int_s^T |u(t+\theta)|^2 dt\nu(d\theta)\nu(d\theta')\Big)^{\frac{1}{2}} \\
&\qquad \times\Big(\int_{[-\delta,0]^2} \int_s^T |u(t+\theta')|^2 dt\nu(d\theta)\nu(d\theta')\Big)^{\frac{1}{2}} \\
&\leqslant K\Big(\int_{-\delta}^0 |\psi(\theta)|^2 d\theta + \int_s^T |u(r)|^2 dr\Big).
\end{aligned}$$

Hereafter, $K > 0$ will be a generic constant which can be different from line to line. Hence by Fubini theorem, we deduce

$$\begin{aligned}
&\int_s^T \int_{[-\delta,0]^2} \big\langle R(t,\theta,\theta')u(t+\theta), u(t+\theta')\big\rangle \nu(d\theta)\nu(d\theta')dt \\
&= \int_{[-\delta,0]^2} \int_s^T \big\langle R(t,\theta,\theta')u(t+\theta), u(t+\theta')\big\rangle dt\nu(d\theta)\nu(d\theta')
\end{aligned}$$

is well-defined. Other terms can be treated similarly. Hence, the cost (1.2.7) is well-defined.

Based on the above observation and analysis, we may consider the state Eq. (1.2.2) with the cost functional (1.2.7). Our optimal control problem can be formulated as follows.

Problem (P) For any given initial tuple $(s, x, \varphi(\cdot), \psi(\cdot))$ from a suitable class, find a control $u^*(\cdot)$ in some suitable class $\mathscr{U}[s, T]$ so that

$$J(s, x, \varphi(\cdot), \psi(\cdot); u^*(\cdot)) = \inf_{u(\cdot)\in\mathscr{U}[s,T]} J(s, x, \varphi(\cdot), \psi(\cdot); u(\cdot)). \tag{1.2.8}$$

If a control $u^*(\cdot)$ exists satisfying (1.2.8), we call it an *open-loop optimal control* of Problem (P), the corresponding state trajectory $X^*(\cdot)$ and the pair $(X^*(\cdot), u^*(\cdot))$ are called the *open-loop optimal state trajectory* associated with $u^*(\cdot)$, and an *open-loop optimal pair* of Problem (P), respectively.

Remark 1.2.3 Our model can cover most cases in the existing literature, for the case with only state delay, readers can refer to [42], [55] (time-invariant coefficients and distributed delays), [31] (time-invariant coefficients, pointwise delays and distributed delays), [26] (time-varying coefficients and pointwise delays), [1], [14], [15], [17], [18], [19], [22], [23] (time-varying coefficients, pointwise delays and distributed delays); for the case with only control delay, readers can refer to [13], [55] (time-invariant coefficients and pointwise delays), [29] (time-invariant coefficients and distributed delays), [38] (time-invariant coefficients, pointwise delays and distributed delays); for the case with both state delay and control delay, readers can refer to [38], [56] (time-invariant coefficients, pointwise delays and distributed delays), [21], [52] (time-invariant coefficients and general delays).

In the end of this section, we consider a control problem in chemical production, and model it as a delayed linear quadratic optimal control problem.

Example 1.2.4 A recycling absorber is a device used in the chemical production process for gas purification or liquid recovery. In this process, the control system needs to adjust the flow rate, pressure, temperature and other parameters of the liquid to ensure the absorption efficiency and product quality. Due to the complexity of physical and chemical processes, control systems may include time delays that may be due to the response time of the sensor, the action time of the actuator, or the transmission time of the process itself.

Consider a simplified cyclic absorber model, and its goal is to keep the concentration of a certain component in the outlet gas stable. For any given $s \in [0, T)$, let $X(\cdot)$ be the concentration of the target component in the outlet gas satisfying the following ODDE:

$$\begin{cases} \dot{X}(t) = -k_1 X(t) + k_2 X(t-\delta) + k_3 u(t) + k_4 u(t-\delta) \\ \qquad\qquad + k_5 f(t-\delta), \quad \text{a.e. } t \in [s, T], \\ X(s) = x, \ X(t) = \varphi(t-s), \quad u(t) = \psi(t-s), \quad t \in [s-\delta, s), \end{cases} \tag{1.2.9}$$

where $u(\cdot)$ is the control input, for example reflux ratio, which can be adjusted to affect absorption efficiency and product quality in actual production. $X(t-\delta)$ is the effect of the concentration at time $t-\delta$ on the concentration change at the current time t. For example, in some chemical processes, the conversion of reactants or the separation of products may not be instantaneous, but take a certain time to reach equilibrium. $f(t-\delta)$ is the concentration of the target ingredient in the feed. Positive constants k_1, k_2, k_3, k_4, k_5 are the rate constants of the different processes in the system.

To maintain the target component concentration in the export gas at the desired level, and make the control cost as small as possible, we consider the following cost functional:

$$J = \int_s^T \Big(q(X(t) - a)^2 + r u(t)^2 \Big) dt, \tag{1.2.10}$$

where a is the desired concentration of the target component in the outlet gas, q and r are weighting factors that quantify the importance of concentration maintenance and the control cost, T is the terminal time, representing scheduled end time of the operation cycle.

Our goal is to find an optimal control input $u^*(\cdot)$ such that (1.2.9) is satisfied and (1.2.10) is minimized, and we denote it **Problem (Ex)**.

1.3 Mathematical Preliminaries

To ensure that this book is self-contained as much as possible, in this section we give some necessary preliminaries.

We first introduce the following definitions and propositions.

Definition 1.3.1 Let E be a real Banach space, $\{\mathcal{S}(t)|t \geqslant 0\} \subseteq \mathscr{L}(E)$. We call $\mathcal{S}(\cdot)$ a C_0-*semigroup* on E if the following holds:

$$\begin{cases} \mathcal{S}(0) = I, \\ \mathcal{S}(t+s) = \mathcal{S}(t)\mathcal{S}(s), \quad \forall\, t, s \geqslant 0, \\ \lim\limits_{s\to 0} |\mathcal{S}(s)x - x|_E = 0, \quad x \in E. \end{cases} \tag{1.3.1}$$

The second and third properties in (1.3.1) are usually referred to as the *semigroup property* and the *strong continuity*, respectively.

Proposition 1.3.2 *Let E be a real Banach space and $\mathcal{S}(\cdot)$ be a C_0-semigroup on E. Then there exist constants $M \geqslant 1$ and $\omega \in \mathbb{R}$, such that*

$$\|\mathcal{S}(t)\|_{\mathscr{L}(E)} \leqslant Me^{\omega t}, \quad t \geqslant 0.$$

Definition 1.3.3 Let E be a real Banach space and $\mathcal{S}(\cdot)$ be a C_0-semigroup on E. Let

$$\begin{cases} \mathscr{D}(A) = \left\{x \in E \mid s\text{-}\lim\limits_{t\to 0} \dfrac{\mathcal{S}(t) - I}{t} x \text{ exists}\right\}, \\ Ax = s\text{-}\lim\limits_{t\to 0} \dfrac{\mathcal{S}(t) - I}{t} x, \qquad \forall\, x \in \mathscr{D}(A), \end{cases}$$

where $s\text{-}\lim$ stands for the strong limit. Operator $A : \mathscr{D}(A) \subseteq E \to E$ is called the *generator* of the C_0-semigroup $\mathcal{S}(\cdot)$. In general, the operator A is unbounded.

Because the generator A uniquely determines the C_0-semigroup $\mathcal{S}(\cdot)$, and in the case $A \in \mathscr{L}(E)$, this C_0-semigroup has an explicit expression $e^{At} \equiv \sum\limits_{k\geqslant 0} \dfrac{A^k t^k}{k!}$, we also denote by e^{At} the C_0-semigroup $\mathcal{S}(\cdot)$ generated by A.

The following results collect some important properties of C_0-semigroup and their generators.

Proposition 1.3.4 *Let E be a real Banach space and e^{At} be a C_0-semigroup on E with A being its generator. Then the following hold.*

(i) *For $k \geqslant 1$, if $x \in \mathscr{D}(A^k)$, then $e^{At}x \in \mathscr{D}(A^k)$ and*

$$A^k e^{At} x = e^{At} A^k x = \frac{d^k}{dt^k}(e^{At}x), \qquad t > 0.$$

(ii) *If $x \in \mathscr{D}(A)$, then*

$$e^{At}x - x = \int_0^t e^{Ar} Axdr = \int_0^t Ae^{Ar}xdr, \qquad t \geqslant 0.$$

(iii) *The generator A is a closed linear operator on E, i.e., the graph $\mathcal{G}(A) \equiv \{(x, Ax) \mid x \in \mathscr{D}(A)\}$ is closed in $E \times E$; and $\bigcap_{k\geqslant 1} \mathscr{D}(A^k)$ is dense in E.*

(iv) *Let $A^\lambda = \lambda A(\lambda I - A)^{-1}$, called the Yosida approximation of A. Then*

$$\lim_{\lambda\to\infty} \|A^\lambda x - Ax\|_E = 0, \quad \forall\, x \in \mathscr{D}(A),$$

$$\lim_{\lambda\to\infty} \sup_{t\in[0,T]} \|e^{A^\lambda t}x - e^{At}x\|_E = 0, \quad \forall\, x \in E,\ T > 0.$$

(v) *The adjoint $(e^{At})^*$ of e^{At} is a semigroup on E^* that is weakly* continuous (as an operator-valued function). In the case where E is reflexive, $(e^{At})^*$ is also a C_0-semigroup on E^* with the generator A^*.*

Next we introduce the following definitions.

Definition 1.3.5 Let E be a real Banach space, and $\Delta_*[0, T] := \{(t, s) \in [0, T]^2 \mid 0 \leqslant s \leqslant t \leqslant T\}$ be the lower triangle domain (including the boundary).

(i) A measurable map $\Phi(t, s) : \Delta_*[0, T] \to \mathscr{L}(E)$ is called a *mild evolution operator*, if $\Phi(\cdot, s)$ is strongly continuous on $[s, T]$ and $\Phi(t, \cdot)$ is strongly continuous on $[0, t]$. Moreover,

$$\begin{cases} \Phi(t,t) = I, \quad t \in [0, T], \\ \Phi(t,r)\Phi(r,s) = \Phi(t,s), \quad 0 \leqslant s \leqslant t \leqslant T. \end{cases}$$

(ii) A *quasi-evolution operator* $\Phi(t, s)$ is a mild evolution operator, such that there exists a closed linear operator valued function $\mathcal{A}(s)$ on E with the dense domain $\mathscr{D}(\mathcal{A}(s)) \subseteq E$, satisfying that there exists at least one non-zero $x \in \mathscr{D}(\mathcal{A}(s))$ for almost all $s \in [0, T]$, the following holds:

$$\Phi(t,s)x - x = \int_s^t \Phi(t,\theta)\mathcal{A}(\theta)xd\theta, \qquad t \in [s, T]. \tag{1.3.2}$$

We denote the set of x satisfying the above by $\mathscr{D}_{\mathcal{A}}$, and we call $\mathcal{A}(\cdot)$ the *quasi-generator* of $\Phi(t, s)$.

(iii) An *almost strong evolution operator* is a mild evolution operator on E, for which there exists an associated closed linear operator $\mathcal{A}(t)$ on E for almost all $t \in [0, T]$, such that

$$\Phi(t, s) : \mathscr{D}(\mathcal{A}(s)) \to \mathscr{D}(\mathcal{A}(t)) \text{ for almost all } t > s \in [0, T], \tag{1.3.3}$$

$$\Phi(t, s)x - x = \int_s^t \mathcal{A}(r)\Phi(r, s)x dr, \quad x \in \mathscr{D}(\mathcal{A}(s)). \tag{1.3.4}$$

Consequently, (1.3.4) implies that

$$\frac{\partial}{\partial t}\Phi(t, s)x = \mathcal{A}(t)\Phi(t, s)x, \quad \text{a.e. for } x \in \mathscr{D}(\mathcal{A}(s)).$$

If $\Phi(t, s)$ satisfies (1.3.3) and (1.3.4) everywhere, then it is called a *strong evolution operator*.

A consequence of the strong continuity of $\Phi(\cdot\,,\cdot)$ is that

$$\operatorname*{esssup}_{\Delta_*[0,T]} \|\Phi(t, s)\|_{\mathscr{L}(E)} < \infty.$$

The following results state some important properties about mild evolution operators and quasi-evolution operators.

Proposition 1.3.6 *Let H be a real Hilbert space.*

(i) *If $\Phi(\cdot\,,\cdot)$ is a mild evolution operator on $\Delta_*[0, T]$ and $D(\cdot) \in L^\infty(0, T; \mathscr{L}(H))$ with*

$$L^\infty(0, T; \mathscr{L}(H)) := \Big\{ D : [0, T] \to \mathscr{L}(H) \;\Big|\; D(\cdot) \text{ is strongly measurable, } \operatorname*{esssup}_{0 \leqslant t \leqslant T} \|D(t)\|_{\mathscr{L}(H)} < \infty \Big\},$$

then the following integral equation

$$\Phi_D(t, s)x = \Phi(t, s)x + \int_s^t \Phi(t, r)D(r)\Phi_D(r, s)x dr$$

has a unique solution $\Phi_D(\cdot\,,\cdot)$ on $\Delta_*[0, T]$ in the class of strongly continuous bounded linear operators on H, and $\Phi_D(\cdot\,,\cdot)$ is a mild evolution operator which is called the perturbed mild evolution operator under the perturbation $D(\cdot)$. Further, if

$$\operatorname*{esssup}_{t \in [0,T]} \|D(t)\|_{\mathscr{L}(H)} \leqslant M_1, \qquad \sup_{(t,s) \in \Delta_*[0,T]} \|\Phi(t, s)\|_{\mathscr{L}(H)} \leqslant M_2,$$

then

$$\|\Phi_D(t,s)\|_{\mathscr{L}(H)} \leqslant M_2 e^{M_1 M_2 (t-s)}, \qquad (t,s) \in \Delta_*[0,T],$$

and $\Phi_D(\cdot\,,\cdot)$ *is also the unique solution to*

$$\Phi_D(t,s)x = \Phi(t,s)x + \int_s^t \Phi_D(t,r)D(r)\Phi(r,s)x dr,$$

in the class of strongly continuous bounded operators on H.

(ii) *If* $\Phi(\cdot\,,\cdot)$ *is a quasi-evolution operator on* H *with quasi-generator* $\mathcal{A}(\cdot)$ *and* $D(\cdot) \in L^\infty(0,T;\mathscr{L}(H))$, *then the perturbed mild evolution operator* $\Phi_D(\cdot\,,\cdot)$ *under perturbation* $D(\cdot)$ *is also a quasi-evolution operator with quasi-generator* $\mathcal{A}(\cdot) + D(\cdot)$.

To conclude this section, we introduce the following definition.

Definition 1.3.7 Let H be a Hilbert space. For $F \in \mathscr{L}(H)$, *a generalized pseudo inverse of* F is defined as a linear operator $F^\dagger : \mathscr{D}(F^\dagger) \to H$ satisfying the following four conditions:

$$FF^\dagger F = F, \quad F^\dagger F F^\dagger = F^\dagger, \quad (FF^\dagger)^* = FF^\dagger, \quad (F^\dagger F)^* = F^\dagger F.$$

When F is injective, $F^\dagger$ is a left inverse of F. When F is surjective, $F^\dagger$ is a right inverse of F. When F is self-adjoint, $F^\dagger$ always exists, which may be unbounded.

1.4 Historic Remarks

The study of LQ optimal control problems can be traced back to the seminal works of Bellman–Glicksberg–Gross in 1958 ([6]), Kalman in 1960 ([39]), and Letov in 1961 ([45]). Since then, the topic attracted the attention of many researchers. Over the past several decades, a vast number of publications have appeared (see, for examples, Anderson–Moore [2], Yong–Zhou [60], Sun–Yong [54], and references cited therein).

Problems in the real world show that many phenomena depend not only on their current states, but also on their past history. Mathematically, this leads to the needs of studying delayed control systems. To our best knowledge, in 1961 the maximum principle for delayed systems was first obtained by Kharatishvili [41] as a necessary condition of optimality. Then in 1962, Krasovskii [42] considered the analytic construction for a delayed LQ optimal control problem and concluded that under certain conditions, a linear control law is optimal. However, the analysis did not provide an explicit characterization of the optimal control. Subsequently there are extensive literature to study the explicit feedback of the optimal control for the delayed LQ control problems. During 1969–1971, Ross–Flügge-Lotz [53], Eller–

Aggarwal–Banks [26], Kushner–Barnea [43], Alekal–Brunovský–Chyung–Lee [1] studied these problems using Carathéodory-Hamilton-Jacobi type arguments. More literature since 1980 can be referred to [11–13, 32, 44, 55] (for time-invariant systems), [35, 50, 57] (for time-varying systems) and references cited therein.

However, the use of the nonreflexive Banach space of continuous function at the level of the classical theory of semigroup, was not completely satisfactory for a purely technical reason. During 1967–1972, people found that for a very large class of delay systems, the state space could be enlarged from continuous functions to an initial point and an initial L^p-function, and this led to the *product space approach*, which was independently introduced by Artola [3–5] for parabolic partial differential equations with delays, Delfour–Mitter [19, 22–25] for nonlinear time-varying delay differential equations, and Borisovic–Turbabin [10] for nonhomogeneous linear time-invariant delay differential equations. Based on the above works, one can derive the feedback of the optimal control for finite dimensional delayed LQ optimal control problems by lifting them to infinite dimensional ones without delays.

For delayed LQ optimal control problems with constant coefficients, usually there are two methods to lift the state: *structural state* and *extended state*. The definition of *structural state* in $\mathbb{R}^n \times L^p$ was introduced by Vinter–Kwong [56] in 1981 to study the infinite time horizon LQ optimal control problem with state and control delays. However, their model only allowed distributed delays in control. The *extended state* in $\mathbb{R}^n \times L^p \times L^p$ was introduced by Ichikawa [38] in 1982, which gave a general theory for a family of evolution equations with a control operator containing a finite number of discrete delays. The structural state roughly corresponds to a transformation of the original state and the segment of the control function by the structural operators associated with the delay structure of the system and the control, while the extended state clearly follows the evolution of the pieces of state trajectory and control. Later in 1984–1986, Delfour [20, 21] generalized the results of [56] to a control operator with a more general delay structure than the one in [38] and to delayed systems with both finite and infinite memories. More literature can be referred to [29–31, 40, 52]. See the book by Bensoussan–Da Prato–Delfour–Mitter [9], Chapter 4, for an excellent survey regarding this aspect.

For delayed time-varying LQ optimal control problems, in 1972, Delfour–Mitter [23] firstly solved an LQ optimal control problem with general state delays by lifting the problem to an abstract one in a suitable Hilbert space, and Delfour [18] later summarized and clarified some results in [23] to present the LQ optimal control problem in the product space framework, which led to infinite dimensional Riccati equations with unbounded operators. In 1975, Curtain [14] introduced the notions of strong evolution operator and mild evolution operator to study infinite-dimensional Riccati equation and its application to LQ problem for time-varying delay systems. Then Curtain–Pritchard [15] extended the results to more general case by introducing quasi-evolution operator, quasi-generator, and (almost) strong evolution operator, etc. However, in the above works ([14, 15]), the control operator is bounded, and thus, the results cannot cover the case of discrete delays in the control. In 1977, Curtain–Pritchard [16] extended the results to the case of unbounded control operator, but without mentioning possible applications in LQ problems with both distributed and discrete delays in both the state and control.

In the literature above, some feedback of the optimal control was represented by lifted operator coefficients, such as [17, 20, 21, 29, 30, 40, 52, 56], while others were explicitly represented by the coefficients of the original delayed control systems, such as [18, 23, 31, 38, 58]. There are also massive research on delay problems about stochastic optimal control, differential game and so on (see [9, 25, 33, 36, 37, 47, 58, 59]).

Recently, Sun–Yong [54] found that there is a significant difference between open-loop and closed-loop saddle points for a stochastic LQ two-person zero-sum differential game. Furthermore, they fundamentally studied the open-loop and closed-loop solvability for general stochastic LQ optimal control problems. Under the convexity of the cost functional, it was established the equivalence between the open-loop solvability and the solvability of the optimality system (which is a forward-backward stochastic differential equations), as well as the equivalence between the strongly regular solvability of the differential Riccati equation and the closed-loop solvability. All these results contain deterministic problems as special cases. See Lü [48, 49] for some further generalization to LQ optimal control problem for systems governed by stochastic evolution equations, even allowing mean-field type terms.

In this book, we consider a general time-varying LQ optimal control Problem (P), in which the state equation and the cost functional both have delays in the state and the control. Moreover, the delays are described by the integral with respect to Radon measures so that the distributed and discrete delays can all be included. Therefore, our model could cover most ones in the literature. We will lift the problem into an LQ problem without delays in a proper infinite dimensional space. Furthermore, we adopt the notions of open-loop solvability, closed-loop solvability, and closed-loop representation of open-loop optimal control for the lifted problem and obtain systematic results relevant to these concepts. Then, we will go back to the original space, representing our final results of optimal controls via the parameters of the original problem (in $\mathbb{R}^n$), completing our presentation.

Radon (signed) measures are quite standard in real analysis. One standard reference is the book by Evans–Gariepy [27] where readers can find some relevant results. The standard theory of C_0-semigroup has been widely used in the study of various problems involving evolution equations. There are many classical references, see, for examples, Hille–Phillips [34], Yosida [61], Pazy [51], see also Curtain–Pritchard [17], Li–Yong [46] and Bensoussan–Da Prato–Delfour–Mitter [9]. As mentioned above, the notions about mild evolution operators were introduced in Curtain–Pritchard [15, 17]. The proof of Proposition 1.3.6 is similar to that of Theorems 9.2 and 9.6 in Curtain–Pritchard [17]. Concerning the generalized pseudo inverse of an operator, readers are referred to Beutler [7, 8], for details.

References

1. Alekal, Y., Brunovský, P., Chyung, D.H., Lee, L.E.: The quadratic problem for systems with time delay. IEEE Trans. Automat. Control **16**, 673–687 (1971)
2. Anderson, B.O.D., Moore, J.B.: Optimal Control: Linear Quadraic Methods. Prentice Hall, Englewood Cliffs (1989)
3. Artola, M.: Équations paraboliques à retardement. C. R. Acad. Sci. Paris Sér. A **264**, 668–671 (1967)
4. Artola, M.: Sur les perturbations des équations d'évolution, application à des problèmes de retard. Ann. Sci. École Norm. Sup. **4**, 137–253 (1969)
5. Artola, M.: Sur une équation du premier ordre à argument retardé. C. R. Acad. Sci. Paris, Śer. A **268**, 1540–1543 (1969)
6. Bellman, R., Glicksberg, I., Gross, O.: Some Aspects of the Mathematical Theory of Control Processes. RAND Corporation, Santa Monica (1958)
7. Beutler, F.J.: The operator theory of the pseudo-inverse I: bounded operators. J. Math. Anal. Appl. **10**, 451–470 (1965)
8. Beutler, F.J.: The operator theory of the pseudo-inverse II: unbounded operators with arbitrary range. J. Math. Anal. Appl. **10**, 471–493 (1965)
9. Bensoussan, A., Da Prato, G., Delfour, M.C., Mitter, S.K.: Representation and Control of Infinite Dimensional Systems, 2nd edn. Birkhäuser, Boston (2007)
10. Borisovic, J.G., Turbabin, A.S.: On the Cauchy problem for linear nonhomogeneous differential equations with retarded argument. Soviet Math. Doklady **10**, 401–405 (1969)
11. Cacace, F., Conte, F., Germani, A.: Memoryless approach to the LQ and LQG problems with variable input delay. IEEE Trans. Automat. Control **61**, 216–221 (2016)
12. Cacace, F., Conte, F., Germani, A., Palombo, G.: Optimal control of linear systems with large and variable input delays. Syst. Conrol Lett. **89**, 1–7 (2016)
13. Carravetta, F., Palumbo, P., Pepe, P.: Memoryless solution to the optimal control problem for linear systems with delayed input. Kybernetika **49**, 568–589 (2013)
14. Curtain, R.F.: The infinite-dimensonal Riccati equation with applications to affine hereditary differential systems. SIAM J. Control Optim. **13**, 1130–1143 (1975)
15. Curtain, R.F., Pritchard, A.J.: The infinite-dimensional Riccati equation for systems defined by the evolution operators. SIAM J. Control Optim. **14**, 951–983 (1976)
16. Curtain, R.F., Pritchard, A.J.: An abstract theory for unbounded control action for distributed parameter systems. SIAM J. Control Optim. **15**, 566–611 (1977)
17. Curtain, R.F., Pritchard, A.J.: Infinite Dimensional Linear Systems Theory, Lecture Notes in Control and Information Sciences, vol. 8. Springer-Verlag, Berlin-New York (1978)
18. Delfour, M.C.: The linear quadratic optimal control problem for hereditary differential systems: theory and numerical solution. Appl. Math. Optim. **3**, 101–162 (1976)
19. Delfour, M.C.: State theory of linear hereditary differential systems. J. Math. Anal. Appl. **60**, 8–35 (1977)
20. Delfour, M.C.: Linear optimal control of systems with state and control variable delays. Automatica J. IFAC **20**, 69–77 (1984)
21. Delfour, M.C.: The linear-quadratic optimal control problem with delays in state and control variables: a state space approach. SIAM J. Control Optim. **24**, 835–883 (1986)
22. Delfour, M.C., Mitter, S.K.: The optimal control of systems governed by affine hereditary differential equations. C. R. Acad. Sci. Paris, Ser. A-B **272**, A1715–A1718 (1971) (in French)
23. Delfour, M.C., Mitter, S.K.: Controllability, observability and optimal feedback control of affine hereditary differential systems. SIAM J. Control Optim. **10**, 298–328 (1972)
24. Delfour, M.C., Mitter, S.K.: Hereditary differential systems with constant delays I. General case. J. Differ. Equ. **12**, 213–235 (1972)
25. Delfour, M.C., Mitter, S.K.: Hereditary differential systems with constant delays II: A class of affine systems and the adjoint problem. J. Differ. Equ. **18**, 18–28 (1975)

26. Eller, D.H., Aggarwal, J.K., Banks, H.T.: Optimal control of linear time-delay systems. IEEE Trans. Automat. Control **AC-14**, 678–687 (1969)
27. Evans, L.C., Gariepy, R.F.: Measure Theory and Fine Properties of Functions. CRC Press, Boca Raton (1992)
28. Fabbri, G., Gozzi, F., Święch, A.: Stochastic Optimal Control in Infinite Dimension: Dynamic Programming and HJB Equations. Springer-Verlag, Berlin (2017). Post-Print
29. Federico, S., Tacconi, E.: Dynamic programming for optimal control problems with delays in the control variable. SIAM J. Control Optim. **52**, 1203–1236 (2014)
30. Flandoli, F.: Solution and control of a bilinear stochastic delay equation. SIAM J. Control Optim. **28**, 934–949 (1990)
31. Gibson, J.S.: Linear-quadratic optimal control of hereditary differential systems: infinite dimensional Riccati equations and numerical approximations. SIAM J. Control Optim. **21**, 95–139 (1983)
32. Grimble, M.J.: The solution of finite-time optimal control problems with control time delays. Optimal Control Appl. Methods **1**, 263–277 (1980)
33. Guatteri, G., Masiero, F.: Stochastic maximum principle for equations with delay: going to infinite dimensions to solve the non-convex case (2023). arXiv:2306.07422
34. Hille, E., Phillips, R.S.: Functional Analysis and Semi-groups. American Mathematical Society, Providence (1982)
35. Huang, J.H., Li, N.: Linear-quadratic mean-field game for stochastic delayed systems. IEEE Trans. Automat. Control **63**, 2722–2729 (2018)
36. Huang, J.H., Shi, J.T.: Maximum principle for optimal control of fully coupled forward-backward stochastic differential delayed equations. ESAIM Control Optim. Calc. Var. **18**, 1073–1096 (2012)
37. Huang, J.H., Li, X., Shi, J.T.: Forward-backward linear quadratic stochastic optimal control problem with delay. Syst. Control Lett. **61**, 623–630 (2012)
38. Ichikawa, A.: Quadratic control of evolution equations with delays in control. SIAM J. Control Optim. **20**, 645–668 (1982)
39. Kalman, R.E.: Contributions to the theory of optimal control. Bol. Soc. Mat. Mexicana **5**, 102–119 (1960)
40. Karrakchou, J., Namir, A.: A new approach for a linear quadratic optimal-control probem of distributed systmes with delays in the control. IMA J. Math Control Inform. **9**, 221–230 (1992)
41. Kharatishvili, G.L.: The maximum principle in the theory of optimal processes involving delay. Dokl. Acad. Nauk SSSR **136**, 39–42 (1961) (in Russian). Transtated as Soviet Math. Dokl., 2 (1961), 28–32
42. Krasovskii, N.N.: On the analytic construction of an optimal control in a system with time lags. J. Appl. Math. Mech. MM **26**, 50–67 (1962)
43. Kushner, H.J., Barnea, D.I.: On the control of a linear functional-differential equation with quadratic cost. SIAM J. Control Optim. **8**, 257–272 (1970)
44. Kwon, W.H., Lee, Y.S., Han, S.H.: General receding horizon control for linear time-delay systems. Automatica J. IFAC **40**, 1603–1611 (2004)
45. Letov, A.M.: The analytical design of control systems. Automat. Remote Control **22**, 363–372 (1961)
46. Li, X., Yong, J.M.: Optimal Control Theory for Infinite Dimensional Systems. Birkhäuser, Boston (1995)
47. Li, N., Wang, G.C., Wu, Z.: Linear-quadratic optimal control for time-delay stochastic system with recursive utility under full and partial information. Automatica J. IFAC **121**, 109169 (2020)
48. Lü, Q.: Well-posedness of stochastic Riccati equations and closed-loop solvability for stochastic linear quadratic optimal control problems. J. Differ. Equ. **267**, 180–227 (2019)
49. Lü, Q.: Stochastic linear quadratic optimal control problems for mean-field stochastic evolution equations. ESAIM Control Optim. Calc. Var. **26**, 127 (2020)

50. Milman, M., Foster, J.H., Schumitzky, A.: Optimal feedback control of infinite dimensional linear systems with applications to hereditary problems. J. Math. Anal. Appl. **119**, 259–281 (1986)
51. Pazy, A.: Semigroups of Linear Operators and Applications to Partial Differential Equations. Springer-Verlag, New York (1983)
52. Pritchard, A.J., Salamon, D.: The linear-quadratic control problem for retarded systens with delays in control and observation. IMA J. Math. Control Inform. **2**, 335–362 (1985)
53. Ross, D.W., Flügge-Lotz, I.: An optimal control problem for systems with differential-difference equation dynamics. SIAM J. Control Optim. **7**, 609–623 (1969)
54. Sun, J.R., Yong, J.M.: Stochastic Linear-Quadratic Optimal Control Theory: Open-Loop and Closed-Loop Solutions. Springer, Berlin (2020)
55. Uchida, K., Shimemura, E.: Closed-loop properties of the infinite-time linear-quadratic optimal regulator for systems with delays. Internat. J. Control **43**, 773–779 (1986)
56. Vinter, R.B., Kwong, R.H.: The infinite time quadratic control problem for linear systems with state and control delays: an evolution equation approach. SIAM J. Control Optim. **19**, 139–153 (1981)
57. Wang, C.M., Jamshidi, M.: Optimal control of large-scale non-linear systems with time-delay. Int. J. Control **39**, 683–699 (1984)
58. Wang, H.X., Zhang, H.S.: LQ control for Itô-type stochastic systems with input delays. Automatica J. IFAC **49**, 3538–3549 (2013)
59. Xu, J.J., Shi, J.T., Zhang, H.S.: A leader-follower stochastic linear quadratic differential game with time delay. Sci. China Inf. Sci. **61**, 112202 (2018)
60. Yong, J.M., Zhou, X.Y.: Stochastic Controls: Hamiltonian Systems and HJB Equations. Springer-Verlag, New York (1999)
61. Yosida, K.: Functional Analysis. Springer-Verlag, Berlin-Heidelberg-New York (1980)

Chapter 2
Problem Lifting

In this chapter, we lift the state equation from $\mathbb{R}^n$ into an infinite dimensional Hilbert space, which transforms the delayed optimal control problem into the one without delays. Some useful basic results will be obtained.

2.1 Abstract Evolution Equations in M^2 Spaces

For ease of reading, we rewrite (1.2.6) as follows.

$$X(t) = x + \int_s^t \int_{(s-\tau,0]} A(\tau,\theta)X(\tau+\theta)\mu(d\theta)d\tau + \int_s^t f(\tau)d\tau, \; t \in [s,T], \tag{2.1.1}$$

where

$$\begin{aligned} f(\tau) = \Big[&\mu(\{s-\tau\})A(\tau,s-\tau)x + \int_{[-\delta,s-\tau)} A(\tau,\theta)\varphi(\tau+\theta-s)\mu(d\theta) \\ &+ \int_{[-\delta,s-\tau)} B(\tau,\theta)\psi(\tau+\theta-s)\mu(d\theta)\Big]\mathbf{1}_{[s,s+\delta]}(\tau) \\ &+ \int_{[s-\tau,0]} B(\tau,\theta)u(\tau+\theta)\mu(d\theta) + b(\tau). \end{aligned} \tag{2.1.2}$$

To study the above equation, let us introduce the following hypothesis.

(H2.1) Let $T \geqslant \delta > 0$ be given. The following two assumptions hold:

(i) $\mu(\cdot)$ is a (scalar) Randon measure on $\mathscr{B}((-\infty,0])$, the Borel σ-field of $(-\infty,0]$, concentrated on $[-\delta,0]$.

© The Author(s), under exclusive license to Springer Nature Singapore Pte Ltd. 2025
W. Meng et al., *Time-Delayed Linear Quadratic Optimal Control Problems*,
SpringerBriefs on PDEs and Data Science,
https://doi.org/10.1007/978-981-96-1897-2_2

(ii) The coefficients $A(\cdot\,,\cdot)$, $B(\cdot\,,\cdot)$ and the nonhomogeneous term $b(\cdot)$ satisfy the following:

$$\begin{cases} A(\cdot\,,\cdot) \in C([0, T+\delta] \times [-\delta, 0]; \mathbb{R}^{n\times n}), \\ B(\cdot\,,\cdot) \in C([0, T+\delta] \times [-\delta, 0]; \mathbb{R}^{n\times m}), \\ b(\cdot) \in L^1(0, T; \mathbb{R}^n). \end{cases}$$

The following theorem gives the well-posedness of the state Eq. (2.1.1), as well as some properties of its solution.

Theorem 2.1.1 *Let* (H2.1) *hold. Then*

(i) *For any* $s \in [0, T]$, $x \in \mathbb{R}^n$, $\varphi(\cdot) \in L^2(-\delta, 0; \mathbb{R}^n)$, $\psi(\cdot) \in L^2(-\delta, 0; \mathbb{R}^m)$ *and* $u(\cdot) \in L^2(s, T; \mathbb{R}^m)$, (2.1.1) *admits a unique solution* $X(\cdot) \equiv X(\cdot\,; s, x, \varphi(\cdot), \psi(\cdot), u(\cdot)) \in C([s, T]; \mathbb{R}^n)$. *Moreover, there exists a constant* $K > 0$ *such that*

$$\begin{aligned} &\|X(\cdot\,; s, x, \varphi(\cdot), \psi(\cdot), u(\cdot))\|_{C([s,T];\mathbb{R}^n)} \\ &\leqslant K\big[|x| + \|\varphi(\cdot)\|_{L^2} + \|\psi(\cdot)\|_{L^2} + \|u(\cdot)\|_{L^2} + \|b(\cdot)\|_{L^1}\big]. \end{aligned} \tag{2.1.3}$$

Furthermore, the following variation of constants formula holds:

$$X(t; s, x, \varphi(\cdot), \psi(\cdot), u(\cdot)) = \Phi^0(t, s)x + \int_s^t \Phi^0(t, r) f(r) dr, \tag{2.1.4}$$

with $f(r)$ *defined by* (2.1.2), *and*

$$\Phi^0(t, s)x := X_0(t; s, x), \quad 0 \leqslant s \leqslant t \leqslant T, \tag{2.1.5}$$

where $X_0(\cdot) \equiv X_0(\cdot\,; s, x)$ *is the solution to the following equation:*

$$\begin{cases} \dot{X}_0(t) = \displaystyle\int_{[s-t,0]} A(t, \theta) X_0(t+\theta) \mu(d\theta), \quad a.e.\ t \in [s, T], \\ X_0(s) = x, \quad X_0(t) = 0, \qquad t \in [s-\delta, s). \end{cases} \tag{2.1.6}$$

(ii) *For any given* $x \in \mathbb{R}^n$, $\varphi(\cdot) \in L^2(-\delta, 0; \mathbb{R}^n)$, $\psi(\cdot) \in L^2(-\delta, 0; \mathbb{R}^m)$ *and* $u(\cdot) \in L^2(0, T; \mathbb{R}^m)$, *the map* $(t, s) \mapsto X(t\,; s, x, \varphi(\cdot), \psi(\cdot), u(\cdot))$ *is continuous on* $\Delta_*[0, T] := \{(t, s) \in [0, T]^2 \mid 0 \leqslant s \leqslant t \leqslant T\}$.

Proof

(i) Let $s \in [0, T)$. For given $x \in \mathbb{R}^n$, $\varphi(\cdot) \in L^2(-\delta, 0; \mathbb{R}^n)$, $\psi(\cdot) \in L^2(-\delta, 0; \mathbb{R}^m)$ and $u(\cdot) \in L^2(s, T; \mathbb{R}^m)$, we recall $f(t)$ as in (2.1.2). Under (H2.1), one has $f(\cdot) \in L^1(s, T; \mathbb{R}^n)$, and

$$\begin{aligned}
&\int_s^T |f(t)|dt \\
&\leqslant K\Big[|x| + \int_s^{(s+\delta)\wedge T} \int_{[-\delta,s-t)} \Big(|\varphi(t+\theta-s)| + |\psi(t+\theta-s)|\Big)\mu(d\theta)dt \\
&\quad + \int_s^T \Big(\int_{[s-t,0]} |u(t+\theta)|\mu(d\theta) + |b(t)|\Big)dt\Big] \\
&= K\Big[|x| + \int_{(-\delta,0]} \int_s^{(s-\theta)\wedge T} \Big(|\varphi(t+\theta-s)| + |\psi(t+\theta-s)|\Big)dt\,\mu(d\theta) \\
&\quad + \int_{[s-t,0]} \int_s^T |u(t+\theta)|dt\,\mu(d\theta) + \int_s^T |b(t)|dt\Big] \\
&\leqslant K\big[|x| + \|\varphi(\cdot)\|_{L^2} + \|\psi(\cdot)\|_{L^2} + \|u(\cdot)\|_{L^2} + \|b(\cdot)\|_{L^1}\big],
\end{aligned} \tag{2.1.7}$$

where, recall that K is a generic constant. Next, for any $s \in [0, T)$, we define a linear operator $\mathscr{A}(s)$ on $C([s, T]; \mathbb{R}^n)$ as follows

$$\mathscr{A}(s)[X(\cdot)](t) := \int_s^t \int_{[s-r,0]} A(r, \theta)X(r+\theta)\mu(d\theta)dr, \qquad t \in [s, T], \quad \forall\, X(\cdot) \in C([s, T]; \mathbb{R}^n).$$

Then

$$\begin{aligned}
&\big\|\mathscr{A}(s)[X(\cdot)]\big\|_{C([s,t];\mathbb{R}^n)} \equiv \sup_{r\in[s,t]} \big|\mathscr{A}(s)[X(\cdot)](r)\big| \\
&\leqslant K_0 \int_s^t \sup_{\tau\in[s,r]} |X(\tau)|dr \equiv K_0 \int_s^t \|X(\cdot)\|_{C([s,r];\mathbb{R}^n)}dr, \quad t \in [s, T],
\end{aligned}$$

with $K_0 > 0$ is an absolute constant. Hence, by induction, we have

$$\big\|\mathscr{A}(s)^k[X(\cdot)]\big\|_{C([s,T];\mathbb{R}^n)} \leqslant \frac{K_0^k(T-s)^k}{k!}\|X(\cdot)\|_{C([s,T];\mathbb{R}^n)}, \quad k \geqslant 1. \tag{2.1.8}$$

Now, we define the following iterated sequence:

$$\begin{cases} X^0(t) = x, & t \in [s, T], \\ X^{k+1}(t) = x + \mathscr{A}(s)[X^k(\cdot)](t) + \displaystyle\int_s^t f(r)dr, & t \in [s, T],\ k \geqslant 1. \end{cases}$$

Clearly, we see that $X^k(\cdot)\in C([s,T];\mathbb{R}^n)$, for each $k\geqslant 1$. Further, by (2.1.8),

$$\begin{aligned}\sup_{\tau\in[s,T]}|X^{k+1}(\tau)-X^k(\tau)| &= \big\|\mathscr{A}(s)^k[X^1(\cdot)-X^0(\cdot)]\big\|_{C([s,T];\mathbb{R}^n)}\\ &\leqslant \frac{K_0^k(T-s)^k}{k!}\|X^1(\cdot)-X^0(\cdot)\|_{C([s,T];\mathbb{R}^n)},\quad k\geqslant 1,\end{aligned}$$

hence $\{X^k(\cdot)\}$ is a Cauchy sequence in $C([s,T];\mathbb{R}^n)$, and its limit is the solution to (2.1.1). To prove the uniqueness, let $\widetilde{X}(\cdot)$ be another solution to (2.1.1), then it is easy to verify that

$$\begin{aligned}\sup_{\tau\in[s,t]}|X(\tau)-\widetilde{X}(\tau)| &\leqslant \int_s^t\int_{[s-r,0]}|A(r,\theta)||X(r+\theta)-\widetilde{X}(r+\theta)|\mu(d\theta)dr\\ &\leqslant K\int_s^t\sup_{\tau\in[s,r]}|X(\tau)-\widetilde{X}(\tau)|dr,\qquad t\in[s,T],\end{aligned}$$

which implies that $\widetilde{X}(\cdot)=X(\cdot)$, by Gronwall's inequality. Further, by a similar argument, we get

$$\sup_{\tau\in[s,t]}|X(\tau)|\leqslant |x|+K\int_s^t\sup_{\tau\in[s,r]}|X(\tau)|dr+\int_s^t|f(r)|dr,\qquad t\in[s,T].$$

Then by Gronwall's inequality again, and combining (2.1.7), we obtain the estimate (2.1.3).

Now, we are proving the variation of constants formula (2.1.4). By (2.1.8), we know that $\sum_{k=0}^{\infty}\|\mathscr{A}(s)^k\|$ is convergent. Thus,

$$X(\cdot)=x+\mathscr{A}(s)[X(\cdot)]+\int_s^{\cdot}f(r)dr=\sum_{k=0}^{\infty}\mathscr{A}(s)^k\Big[x+\int_s^{\cdot}f(r)dr\Big].$$

By letting

$$\Phi^0(t,s)x=\sum_{k=0}^{\infty}\mathscr{A}(s)^k[x](t)=x+\mathscr{A}(s)\big[\Phi^0(\cdot,s)x\big](t),$$

we have

$$X(t)=\Phi^0(t,s)x+\sum_{k=0}^{\infty}\mathscr{A}(s)^k\Big[\int_s^{\cdot}f(r)dr\Big](t).$$

Note that

$$\Phi^0(t,s)x = x + \int_s^t \int_{[s-r,0]} A(r,\theta)\Phi^0(r+\theta,s)x\mu(d\theta)dr, \qquad t\in[s,T],$$

then we get (2.1.5)–(2.1.6). Moreover, noting

$$\begin{aligned}&\int_s^t \int_\tau^t \int_{[\tau-r,0]} A(r,\theta)\Phi^0(r+\theta,\tau)f(\tau)\mu(d\theta)drd\tau\\ &= \int_s^t \int_{[s-r,0]} A(r,\theta)\int_s^{r+\theta} \Phi^0(r+\theta,\tau)f(\tau)d\tau\mu(d\theta)dr,\end{aligned}$$

we obtain

$$\int_s^t \Phi^0(t,\tau)f(\tau)d\tau = \int_s^t f(\tau)d\tau + \mathscr{A}(s)\Big[\int_s^\cdot \Phi^0(\cdot,\tau)f(\tau)d\tau\Big](t).$$

Thus we deduce

$$\begin{aligned}&\sum_{k=0}^\infty \mathscr{A}(s)^k\Big[\int_s^\cdot f(r)dr\Big](t) = \int_s^t f(r)dr + \mathscr{A}(s)\Big[\int_s^\cdot \Phi^0(\cdot,\tau)f(\tau)d\tau\Big](t)\\ &= \int_s^t f(r)dr + \int_s^t \int_{[s-r,0]} A(r,\theta)\Big[\int_s^{r+\theta} \Phi^0(r+\theta,\tau)f(\tau)d\tau\Big]\mu(d\theta)dr\\ &= \int_s^t f(\tau)d\tau + \int_s^t \Big[\int_\tau^t \int_{[\tau-r,0]} A(r,\theta)\Phi^0(r+\theta,\tau)\mu(d\theta)dr\Big]f(\tau)d\tau\\ &= \int_s^t \Big[I + \int_\tau^t \int_{[\tau-r,0]} A(r,\theta)\Phi^0(r+\theta,\tau)\mu(d\theta)dr\Big]f(\tau)d\tau\\ &= \int_s^t \Phi^0(t,\tau)f(\tau)d\tau.\end{aligned}$$

This proves (2.1.4).

(ii) Next, we prove that for any given $(t,x,\varphi(\cdot),\psi(\cdot),u(\cdot))$, $s\mapsto X(t;s,x,\varphi(\cdot),\psi(\cdot),u(\cdot))$ is continuous. Without loss of generality, we let $s<s'<s'+\delta<t$, and denote $X(t) = X(t;s,x,\varphi(\cdot),\psi(\cdot),u(\cdot))$, $X'(t) = X(t;s',x,\varphi(\cdot),\psi(\cdot),u(\cdot))$. Then from (2.1.1) and by Fubini theorem,

$$\begin{aligned}X(t) = x &+ \int_{[-\delta,0]} \int_{s-\theta}^t \Big[A(r,\theta)X(r+\theta) + B(r,\theta)u(r+\theta)\Big]dr\mu(d\theta)\\ &+ \int_{[-\delta,0]} \int_s^{s-\theta} \Big[A(r,\theta)\varphi(r+\theta-s) + B(r,\theta)\psi(r+\theta-s)\Big]dr\mu(d\theta)\\ &+ \int_s^t b(r)dr, \qquad t\in[s,T],\end{aligned}$$

$$
\begin{aligned}
X'(t) = x &+ \int_{[-\delta,0]} \int_{s'-\theta}^{t} \Big[A(r,\theta)X'(r+\theta) + B(r,\theta)u(r+\theta)\Big] dr\,\mu(d\theta) \\
&+ \int_{[-\delta,0]} \int_{s'}^{s'-\theta} \Big[A(r,\theta)\varphi(r+\theta-s') \\
&+ B(r,\theta)\psi(r+\theta-s')\Big] dr\,\mu(d\theta) \\
&+ \int_{s'}^{t} b(r)dr, \qquad t \in [s', T].
\end{aligned}
$$

Thus,

$$
X'(t) - X(t) = \int_{[-\delta,0]} \int_{s'-\theta}^{t} A(r,\theta)(X'(r+\theta) - X(r+\theta)) dr\,\mu(d\theta) + g(s,s'), \tag{2.1.9}
$$

with

$$
\begin{aligned}
g(s,s') = &- \int_{[-\delta,0]} \int_{s-\theta}^{s'-\theta} \big(A(r,\theta)X(r+\theta) + B(r,\theta)u(r+\theta)\big) dr\,\mu(d\theta) \\
&+ \int_{[-\delta,0]} \int_{\theta}^{0} \big[\big(A(r+s'-\theta,\theta) - A(r+s-\theta,\theta)\big)\varphi(r) \\
&+ \big(B(r+s'-\theta,\theta) - B(r+s-\theta,\theta)\big)\psi(r)\big] dr\,\mu(d\theta) - \int_{s}^{s'} b(r)dr.
\end{aligned}
$$

Note that

$$
\begin{aligned}
&\Big|\int_{[-\delta,0]} \int_{s-\theta}^{s'-\theta} \big(A(r,\theta)X(r+\theta) + B(r,\theta)u(r+\theta)\big) dr\,\mu(d\theta)\Big| \\
&\leqslant K \int_{s}^{s'} \big(|X(r)| + |u(r)|\big) dr,
\end{aligned}
$$

and

$$
\begin{aligned}
&\Big|\int_{[-\delta,0]} \int_{\theta}^{0} \big[\big(A(r+s'-\theta,\theta) - A(r+s-\theta,\theta)\big)\varphi(r) \\
&\qquad + \big(B(r+s'-\theta,\theta) - B(r+s-\theta,\theta)\big)\psi(r)\big] dr\,\mu(d\theta)\Big| \\
&= \Big|\int_{[-\delta,0]^2} \Big[\big(A(r+s'-\theta,\theta) - A(r+s-\theta,\theta)\big)\varphi(r)\mathbf{1}_{[\theta,0]}(r) \\
&\quad + \big(B(r+s'-\theta,\theta) - B(r+s-\theta,\theta)\big)\psi(r)\mathbf{1}_{[\theta,0]}(r)\Big] dr\,\mu(d\theta)\Big|
\end{aligned}
$$

$$
\begin{aligned}
&\leqslant \Big(\int_{[-\delta,0]^2} |\varphi(r)|^2 \mathbf{1}_{[\theta,0]}(r) dr\, \mu(d\theta)\Big)^{\frac{1}{2}} \\
&\quad \times \Big(\int_{[-\delta,0]^2} |A(r+s'-\theta,\theta)-A(r+s-\theta,\theta)|^2 \mathbf{1}_{[\theta,0]}(r) dr\, \mu(d\theta)\Big)^{\frac{1}{2}} \\
&\quad + \Big(\int_{[-\delta,0]^2} |\psi(r)|^2 \mathbf{1}_{[\theta,0]}(r) dr\, \mu(d\theta)\Big)^{\frac{1}{2}} \\
&\quad \times \Big(\int_{[-\delta,0]^2} |B(r+s'-\theta,\theta)-B(r+s-\theta,\theta)|^2 \mathbf{1}_{[\theta,0]}(r) dr\, \mu(d\theta)\Big)^{\frac{1}{2}} \\
&\leqslant \Big(\|\varphi\|_{L^2} + \|\psi\|_{L^2}\Big)\sqrt{\mu([-\delta,0])}\, \omega(|s-s'|),
\end{aligned}
$$

where $\omega(\cdot)$ is the modulus of continuity for $A(\cdot\,,\cdot)$ and $B(\cdot\,,\cdot)$. Hence, (2.1.9) implies

$$
|X'(t) - X(t)| \leqslant K \int_{s'}^{t} |X'(r) - X(r)| dr + |g(s,s')|,
$$

by Gronwall's inequality we deduce

$$
|X'(t) - X(t)| \leqslant |g(s,s')| e^{Kt} \to 0, \ \text{as } |s'-s| \to 0.
$$

Therefore, $X(t; s, x, \varphi(\cdot), \psi(\cdot), u(\cdot))$ is continuous in $s \in [0,t]$ and similarly it is also continuous in t on $[s,T]$. Thus we complete the proof.

□

Next, we define $M^2 := \mathbb{R}^n \times L^2(-\delta, 0; \mathbb{R}^n)$, with the inner product

$$
\langle \xi, \xi' \rangle_{M^2} := \langle x, x' \rangle + \int_{-\delta}^{0} \langle \varphi(\theta), \varphi'(\theta) \rangle d\theta,
$$

and

$$
\|\xi\|_{M^2} := \langle \xi, \xi \rangle_{M^2}^{\frac{1}{2}}, \quad \forall\, \xi = \begin{pmatrix} x \\ \varphi \end{pmatrix}, \ \xi' = \begin{pmatrix} x' \\ \varphi' \end{pmatrix} \in M^2.
$$

In what follows, sometimes, we also denote $\xi = \begin{pmatrix} \xi^0 \\ \xi^1 \end{pmatrix}$ with $\xi^0 \in \mathbb{R}^n$ and $\xi^1 \in L^2(-\delta, 0; \mathbb{R}^n)$. Define the mapping $\Phi(\cdot\,,\cdot)$ as follows:

$$
\Phi(t,s) : M^2 \longrightarrow M^2, \quad \xi \longmapsto \begin{pmatrix} \mathfrak{X}(t) \\ \mathfrak{X}_t(\cdot) \end{pmatrix}, \quad \forall\, \xi := \begin{pmatrix} x \\ \varphi \end{pmatrix} \in M^2, \tag{2.1.10}
$$

where $\mathfrak{X}(\cdot) \equiv \mathfrak{X}(\cdot\,; s, x, \varphi(\cdot))$ is the solution to (2.1.1) with $B(\cdot\,,\cdot) \equiv 0$ and $b(\cdot) \equiv 0$, $\mathfrak{X}_t(\theta) := \mathfrak{X}(t+\theta)$, $\theta \in [-\delta, 0]$, i.e.,

$$\begin{aligned}\mathfrak{X}(t) = x + &\int_s^t \int_{(s-r,0]} A(r,\theta)\mathfrak{X}(r+\theta)\mu(d\theta)dr \\ &+\int_s^{t\wedge(s+\delta)} \Big[\mu(\{s-r\})A(r,s-r)x + \int_{[-\delta,s-r)} A(r,\theta)\varphi(r+\theta-s)\mu(d\theta)\Big]dr.\end{aligned} \tag{2.1.11}$$

Theorem 2.1.2 *Let* (H2.1) *hold. Then*

(i) $\Phi(\cdot\,,\cdot) : \Delta_*[0,T] \to \mathscr{L}(M^2)$ *is a mild evolution operator.*

(ii) *Given* $\xi(\cdot) = \begin{pmatrix} x \\ \varphi(\cdot)\end{pmatrix} \in M^2$, $\psi(\cdot) \in L^2(-\delta, 0; \mathbb{R}^m)$, $u(\cdot) \in L^2(0,T;\mathbb{R}^m)$, *denote* $\mathbf{X}(t; s, \xi, \psi, u) := \begin{pmatrix} X(t; s, \xi, \psi, u) \\ X_t(\cdot\,; s, \xi, \psi, u)\end{pmatrix}$, *where* $X(\cdot\,; s, \xi, \psi, u)$ *is the solution to* (2.1.1)*, then*

$$\mathbf{X}(t; s, \xi, \psi, u) = \Phi(t,s)\xi + \int_s^t \Phi(t,r)\Big[\widetilde{B}(r)u_r + \tilde{b}(r)\Big]dr, \quad t \in [s,T], \tag{2.1.12}$$

where $u_r(\theta) := u(r+\theta)$, $\theta \in [-\delta, 0]$, $\tilde{b}(r) := \begin{pmatrix} b(r) \\ 0 \end{pmatrix}$, *and*

$$\widetilde{B}(r) : L^2(-\delta, 0; \mathbb{R}^m) \longrightarrow M^2; \quad \psi(\cdot) \mapsto \begin{pmatrix} \int_{[-\delta,0]} B(r,\theta)\psi(\theta)\mu(d\theta) \\ 0 \end{pmatrix}. \tag{2.1.13}$$

Furthermore, there exists a constant $K > 0$ *such that*

$$\begin{aligned}&\|\mathbf{X}(\cdot\,; s, \xi, \psi, u)\|_{C([s,T];M^2)} \\ &\leqslant K\Big[\|\xi(\cdot)\|_{M^2} + \|\psi(\cdot)\|_{L^2} + \|u(\cdot)\|_{L^2} + \|b(\cdot)\|_{L^1}\Big].\end{aligned} \tag{2.1.14}$$

Proof

(i) $\Phi(t,s) \in \mathscr{L}(M^2)$ can be verified by Theorem 2.1.1 (i), and it is also easy to verify that for any $\xi \in M^2$,

$$\begin{cases} \Phi(t,t)\xi = \xi, & t \in [0,T], \\ \Phi(t,r)\Phi(r,s)\xi = \Phi(t,s)\xi, & 0 \leqslant s \leqslant r \leqslant t \leqslant T. \end{cases}$$

Thus it remains to prove that $\Phi(\cdot, s)$ is strongly continuous on $[s, T]$ and $\Phi(t, \cdot)$ is strongly continuous on $[0, t]$, which can be obtained by Theorem 2.1.1 (ii).

(ii) First the right-hand side of (2.1.12) is well-defined. In fact, by the definition (2.1.5) of $\Phi^0(\cdot, \cdot)$,

$$\left[\Phi(t,r)\begin{pmatrix} h^0 \\ 0 \end{pmatrix}\right]^0 = \Phi^0(t,r)h^0, \tag{2.1.15}$$

where $\left[\Phi(t,r)\begin{pmatrix} h^0 \\ 0 \end{pmatrix}\right]^0$ is the component in $\mathbb{R}^n$ of $\left[\Phi(t,r)\begin{pmatrix} h^0 \\ 0 \end{pmatrix}\right]$. By the continuity of $B(\cdot\,,\cdot)$ and the strong continuity of $\Phi^0(\cdot\,,\cdot)$, one sees that $\int_s^t \Phi(t,r)\widetilde{B}(r)u_r dr$ can be understood as follows:

$$\begin{aligned}
\int_s^t \Phi(t,r)\widetilde{B}(r)u_r dr &= \int_s^t \Phi(t,r)\begin{pmatrix} \int_{[-\delta,0]} B(r,\theta)u(r+\theta)\mu(d\theta) \\ 0 \end{pmatrix} dr \\
&= \begin{pmatrix} \int_s^t \Phi^0(t,r)\int_{[-\delta,0]} B(r,\theta)u(r+\theta)\mu(d\theta)dr \\ \int_s^t \Phi^0(t+\cdot,r)\int_{[-\delta,0]} B(r,\theta)u(r+\theta)\mu(d\theta)dr \end{pmatrix} \\
&= \begin{pmatrix} \int_s^t \int_{[-\delta,0]} \Phi^0(t,r)B(r,\theta)u(r+\theta)\mu(d\theta)dr \\ \int_s^t \int_{[-\delta,0]} \Phi^0(t+\cdot,r)B(r,\theta)u(r+\theta)\mu(d\theta)dr \end{pmatrix} \\
&= \begin{pmatrix} \int_{[-\delta,0]} \int_{s+\theta}^{t+\theta} \Phi^0(t,r-\theta)B(r-\theta,\theta)u(r)dr\mu(d\theta) \\ \int_{[-\delta,0]} \int_{s+\theta}^{t+\theta} \Phi^0(t+\cdot,r-\theta)B(r-\theta,\theta)u(r)dr\mu(d\theta) \end{pmatrix},
\end{aligned} \tag{2.1.16}$$

where the third equality holds by the definition (2.1.5) of $\Phi^0(\cdot\,,\cdot)$.

Now, we consider the following two equations:

$$\begin{cases} \dot{X}^1(t) = \int_{[-\delta,0]} A(t,\theta)X^1(t+\theta)\mu(d\theta), \quad \text{a.e. } t \in [s,T], \\ X^1(s) = x, \quad X^1(t) = \varphi(t-s), \ t \in [s-\delta, s), \end{cases}$$

and

$$\begin{cases} \dot{X}^2(t) = \int_{[-\delta,0]} \Big[A(t,\theta)X^2(t+\theta) + B(t,\theta)u(t+\theta)\Big]\mu(d\theta) + b(t), \\ \qquad\qquad\qquad\qquad\qquad\qquad \text{a.e. } t \in [s,T], \\ X^2(t) = 0, \ t \in [s-\delta, s], \quad u(t) = \psi(t-s), \ t \in [s-\delta, s). \end{cases}$$

Apparently, $X(t; s, \xi, \psi, u) = X^1(t) + X^2(t)$, $t \in [s - \delta, T]$. Denote

$$\mathbf{X}^1(t) = \begin{pmatrix} X^1(t) \\ X^1_t(\cdot) \end{pmatrix}, \quad \mathbf{X}^2(t) = \begin{pmatrix} X^2(t) \\ X^2_t(\cdot) \end{pmatrix},$$

then $\mathbf{X}(t; s, \xi, \psi, u) = \mathbf{X}^1(t) + \mathbf{X}^2(t)$. Recalling $\mathbf{X}^1(t) = \Phi(t, s)\xi$, it remains to prove that

$$\mathbf{X}^2(t) = \int_s^t \Phi(t, r)\Big[\widetilde{B}(r)u_r + \tilde{b}(r)\Big]dr, \quad t \in [s, T]. \tag{2.1.17}$$

From (2.1.15), we have

$$\Big\{\Phi(t, r)\Big[\widetilde{B}(r)u_r + \tilde{b}(r)\Big]\Big\}^0 = \Phi^0(t, r)\Big[\int_{[-\delta,0]} B(r, \theta)u(r + \theta)\mu(d\theta) + b(r)\Big],$$

where $\{\Phi(t, r)[\widetilde{B}(r)u_r + \tilde{b}(r)]\}^0$ is the component in $\mathbb{R}^n$ of $\Phi(t, r)[\widetilde{B}(r)u_r + \tilde{b}(r)]$. By Theorem 2.1.1 (i), we get

$$\begin{aligned} X^2(t) &= \int_s^t \Phi^0(t, r)\Big[\int_{[-\delta,0]} B(r, \theta)u(r + \theta)\mu(d\theta) + b(r)\Big]dr \\ &= \Big\{\int_s^t \Phi(t, r)\Big[\widetilde{B}(r)u_r + \tilde{b}(r)\Big]dr\Big\}^0. \end{aligned} \tag{2.1.18}$$

Noting

$$\left[\Phi(t, r)\begin{pmatrix} h^0 \\ 0 \end{pmatrix}\right]^1(\theta) = \begin{cases} \Phi^0(t + \theta, r)h^0, & \text{if } t + \theta \geqslant r; \\ 0, & \text{if } t + \theta < r, \end{cases}$$

which and (2.1.18) imply that

$$\begin{aligned} &\int_s^t \Big[\Phi(t, r)[\widetilde{B}(r)u_r + \tilde{b}(r)]\Big]^1(\theta)dr \\ &= \int_s^{t+\theta} \Phi^0(t + \theta, r)\Big[\int_{[-\delta,0]} B(r, \beta)u(r + \beta)\mu(d\beta) + b(r)\Big]dr = X^2(t + \theta). \end{aligned}$$

Thus, we have

$$X^2(t + \theta) = \Big[\int_s^t \Phi(t, r)[\widetilde{B}(r)u_r + \tilde{b}(r)]dr\Big]^1(\theta), \quad \text{a.e. } \theta \in [-\delta, 0],$$

which and (2.1.18) imply that (2.1.17) holds. Finally by the strong continuity of $\Phi(\cdot, s)$ on $[s, T]$, $\mathbf{X}(\cdot; s, \xi, \psi, u) \in C([s, T]; M^2)$, furthermore by (2.1.3), (2.1.14) can be proved easily, thus we complete the proof of Theorem 2.1.2.

□

By Theorem 2.1.2, the original state Eq. (2.1.1) is equivalent to (2.1.12), where the state delays disappear, next we deal with the control delays.

Denote $L^2 := L^2(-\delta, 0; \mathbb{R}^m)$ and introduce the following semigroup of left translation:

$$e^{Dt}(t) : L^2 \to L^2, \quad \big[e^{Dt}(t)Y\big](\theta) := \begin{cases} \begin{cases} Y(t+\theta), & -\delta \leqslant \theta \leqslant -t, \\ 0, & -t < \theta \leqslant 0, \end{cases} & \text{if } t \leqslant \delta, \\ 0, \quad\quad -\delta \leqslant \theta \leqslant 0, & \text{if } t > \delta, \end{cases}$$

its generator is given by

$$D : W \longrightarrow L^2; \quad DY := \frac{dY}{d\theta}, \quad \forall\, Y \in W, \tag{2.1.19}$$

where

$$W := \mathscr{D}(D) = \Big\{ Y \in H^1(-\delta, 0; \mathbb{R}^m) \;\Big|\; Y \text{ is absolutely continuous}, \;\; Y(0) = 0 \Big\}.$$

Consider the following abstract differential equation in L^2 space:

$$\begin{cases} \dfrac{d\mathbf{Y}_t}{dt} = D\Big(\mathbf{Y}_t - Fu(t)\Big), \quad t \in [s, T], \\ \mathbf{Y}_s = \psi \in L^2, \end{cases} \tag{2.1.20}$$

where $F \in \mathscr{L}(\mathbb{R}^m, L^2)$ is defined by $\big[F(u)\big](\theta) := u, \; -\delta \leqslant \theta \leqslant 0, u \in \mathbb{R}^m$. The mild solution to (2.1.20) is defined as follows

$$\mathbf{Y}_t = e^{D(t-s)}\psi - D\int_s^t e^{D(t-r)} Fu(r)dr, \quad t \in [s, T]. \tag{2.1.21}$$

Lemma 2.1.3 *The equality* (2.1.21) *is equivalent to*

$$\mathbf{Y}_t(\theta) = \begin{cases} u(t+\theta), & t+\theta \in [s, T], \\ \psi(\theta + t - s), & t+\theta \in [s-\delta, s). \end{cases} \tag{2.1.22}$$

Thus $\mathbf{Y}_t(\theta) = u_t(\theta)$, $t \in [s, T]$, $\theta \in [-\delta, 0]$.

Proof By the definition of the operator F, for any given $r \in [s, T]$, we have

$$[Fu(r)](\theta) = u(r), \quad -\delta \leqslant \theta \leqslant 0.$$

For any given $t \in [s, T]$, by the definition of e^{Dt}, we get

$$\Big(\int_s^t e^{D(t-r)} Fu(r)dr\Big)(\theta) = \int_{s \vee (t+\theta)}^t u(r)dr, \quad \theta \in [-\delta, 0],$$

thus $\left(\int_s^t e^{D(t-r)}Fu(r)dr\right)(\cdot) \in W$, and

$$\begin{aligned}\left(D\int_s^t e^{D(t-r)}Fu(r)dr\right)(\theta) &= -u(t+\theta)\mathbf{1}_{(s-t,0)}(\theta)\\ &= \begin{cases}\begin{cases}0, & -\delta \leqslant \theta \leqslant s-t,\\ -u(t+\theta), & s-t<\theta\leqslant 0,\end{cases} & \text{if } t-s\leqslant\delta,\\ -u(t+\theta), \quad -\delta\leqslant\theta\leqslant 0, & \text{if } t-s>\delta.\end{cases}\end{aligned} \tag{2.1.23}$$

By the definition of e^{Dt}, we obtain

$$\left[e^{D(t-s)}\psi\right](\theta) = \begin{cases}\begin{cases}\psi(t-s+\theta), & -\delta\leqslant\theta\leqslant s-t,\\ 0, & s-t<\theta\leqslant 0,\end{cases} & \text{if } t-s\leqslant\delta,\\ 0, \quad -\delta\leqslant\theta\leqslant 0, & \text{if } t-s>\delta.\end{cases} \tag{2.1.24}$$

Combining (2.1.21), (2.1.23) and (2.1.24), we deduce

$$\mathbf{Y}_t(\theta) = \begin{cases}\begin{cases}\psi(t-s+\theta), & -\delta\leqslant\theta\leqslant s-t,\\ u(t+\theta), & s-t<\theta\leqslant 0,\end{cases} & \text{if } t-s\leqslant\delta,\\ u(t+\theta), \quad -\delta\leqslant\theta\leqslant 0, & \text{if } t-s>\delta,\end{cases}$$

which implies (2.1.22). □

By Lemma 2.1.3, we can replace the control delays u_t with the new variables $\mathbf{Y}_t$. Introduce the following bounded linear operator:

$$\mathcal{D}:\mathbb{R}^m \longrightarrow V^*, \quad \langle\mathcal{D}u, w\rangle_{V^*,V} := \langle u, w(0)\rangle, \quad \forall\, u\in\mathbb{R}^m,\ w\in V, \tag{2.1.25}$$

where $V := H^1(-\delta, 0; \mathbb{R}^m)$, V^* is the dual of V. The following lemma implies that $\int_s^t e^{D(t-r)}\mathcal{D}u(r)dr$ can be regarded as a function in L^2.

Lemma 2.1.4 *For each $u(\cdot)\in L^2(s, T; \mathbb{R}^m)$ and $\alpha \geqslant s$,*

$$\begin{aligned}-D\int_\alpha^t e^{D(t-r)}Fu(r)dr &= \int_\alpha^t e^{D(t-r)}\mathcal{D}u(r)dr\\ &= \begin{cases}u(t+\theta), \quad -\delta\leqslant\theta\leqslant 0, & \text{if } t-\alpha>\delta,\\ \begin{cases}u(t+\theta), & \alpha-t<\theta\leqslant 0,\\ 0, & -\delta\leqslant\theta\leqslant\alpha-t,\end{cases} & \text{if } t-\alpha\leqslant\delta.\end{cases}\end{aligned}$$

Proof Denote by e^{D^*t} the adjoint operator of e^{Dt}, then we have

$$[e^{D^*(t-r)}Y](\theta) = \begin{cases}0, \quad -\delta\leqslant\theta\leqslant 0, & \text{if } t-r>\delta,\\ \begin{cases}Y(r+\theta-t), & t-r-\delta\leqslant\theta\leqslant 0,\\ 0, & -\delta\leqslant\theta<t-r-\delta,\end{cases} & \text{if } t-r\leqslant\delta.\end{cases}$$

For any $w \in V$, by the definition of adjoint operators on normed linear spaces, we have

$$\begin{aligned}
&\Big\langle \int_\alpha^t e^{D(t-r)}\mathcal{D}u(r)dr, w\Big\rangle_{V^*,V} = \int_\alpha^t \big\langle e^{D(t-r)}\mathcal{D}u(r), w\big\rangle_{V^*,V}dr \\
&= \int_\alpha^t \big\langle \mathcal{D}u(r), e^{D^*(t-r)}w\big\rangle_{V^*,V}dr = \int_\alpha^t \big\langle u(r), \big(e^{D^*(t-r)}w\big)(0)\big\rangle dr \\
&= \int_{\alpha\vee(t-\delta)}^t \big\langle w(r-t), u(r)\big\rangle dr = \int_{-\delta}^0 \big\langle w(\theta), u(t+\theta)\mathbf{1}_{(\alpha-t,\infty)}(\theta)\big\rangle d\theta,
\end{aligned}$$

thus we deduce

$$\Big[\int_\alpha^t e^{D(t-r)}\mathcal{D}u(r)dr\Big](\theta) = \begin{cases} u(t+\theta), & -\delta \leqslant \theta \leqslant 0, \qquad \text{if } t-\alpha > \delta, \\ \begin{cases} u(t+\theta), & \alpha - t < \theta \leqslant 0, \\ 0, & -\delta \leqslant \theta \leqslant \alpha - t, \end{cases} & \text{if } t-\alpha \leqslant \delta, \end{cases}$$

which and (2.1.23) complete the proof of Lemma 2.1.4. □

By Lemma 2.1.4, (2.1.21) can be written as follows:

$$\mathbf{Y}_t = e^{D(t-s)}\psi + \int_s^t e^{D(t-r)}\mathcal{D}u(r)dr, \quad t \in [s, T]. \tag{2.1.26}$$

Therefore, by denoting $\mathbf{X}(\cdot) \equiv \mathbf{X}(\cdot\,; s, \xi, \psi, u)$, we can rewrite (2.1.12) as follows

$$\begin{cases} \mathbf{X}(t) = \Phi(t,s)\xi + \int_s^t \Phi(t,r)\big[\widetilde{B}(r)\mathbf{Y}_r + \tilde{b}(r)\big]dr, & t \in [s, T], \\ \mathbf{Y}_t = e^{D(t-s)}\psi + \int_s^t e^{D(t-r)}\mathcal{D}u(r)dr, & t \in [s, T], \end{cases} \tag{2.1.27}$$

which is equivalent to (2.1.1). In (2.1.27), the control delays disappear.

Finally, we try to merge the two equations of (2.1.27) into one equation and regard it as the new state equation of a lifted infinite dimensional optimal control problem. Denote $Z := M^2 \times L^2$ with the inner product:

$$\langle z_1, z_2\rangle_Z := \langle \xi_1, \xi_2\rangle_{M^2} + \langle \psi_1, \psi_2\rangle_{L^2}, \qquad \forall z_1 = \begin{pmatrix}\xi_1\\ \psi_1\end{pmatrix}, z_2 = \begin{pmatrix}\xi_2\\ \psi_2\end{pmatrix} \in Z.$$

Define the following mild evolution operator:

$$\begin{aligned}
&\mathbf{T}(t,s) : Z \longrightarrow Z \\
&\mathbf{T}(t,s)\begin{pmatrix}\xi\\ \psi\end{pmatrix} := \begin{bmatrix} \Phi(t,s)\xi + \int_s^t \Phi(t,r)\widetilde{B}(r)e^{D(r-s)}\psi dr \\ varepsilon^{D(t-s)}\psi \end{bmatrix},
\end{aligned} \tag{2.1.28}$$

and $\mathbf{Z}_0 := \begin{pmatrix} \xi \\ \psi \end{pmatrix}$, $\mathbf{Z}(\cdot) := \begin{pmatrix} \mathbf{X}(\cdot) \\ \mathbf{Y}. \end{pmatrix}$, $\mathbf{b}(t) := \begin{pmatrix} \tilde{b}(t) \\ 0 \end{pmatrix}$, where the integral $\int_s^t \Phi(t,r)\widetilde{B}(r)e^{D(r-s)}\psi dr$ can be understood in a way similar to (2.1.16). Then (2.1.27) can be written as follows

$$\mathbf{Z}(t) = \mathbf{T}(t,s)\mathbf{Z}_0 + \int_s^t \mathbf{T}(t,r)\big[\mathbf{B}u(r) + \mathbf{b}(r)\big]dr, \tag{2.1.29}$$

where $\mathbf{B} := \begin{pmatrix} 0 \\ \mathcal{D} \end{pmatrix}$, it maps $\mathbb{R}^m$ to $M^2 \times V^*$ out Z and thus $\mathbf{B} \notin \mathscr{L}(\mathbb{R}^m, Z)$. Notice that the integral $\int_s^t \mathbf{T}(t,r)\mathbf{B}u(r)dr$ is not defined in Z, (2.1.29) is just a formal expression and actually means (2.1.27).

Now by lifting the original state Eq. (2.1.1), we obtain the new state Eq. (2.1.29) (or (2.1.27)), which contains neither the state delays nor the control delays. However, (2.1.29) is not a standard state equation of an infinite dimensional optimal control problem, and we have to use some new methods to deal with it, different from ways in [4, 14–16].

2.2 Abstract LQ Problems

In this section, we will lift the cost functional and then the delayed Problem (P) can be transformed into an equivalent infinite dimensional control problem without delays. Finally we give the definitions of the open-loop and closed-loop solvability for Problem (P).

Now we would like to rewrite the cost functional (1.2.7) in terms of $\mathbf{Z}(\cdot)$ (or $\mathbf{X}(\cdot)$ and $\mathbf{Y}.$) and $u(\cdot)$, recall the cost (1.2.7) taking the following form:

$$\begin{aligned} &J(s,x,\varphi(\cdot),\psi(\cdot);u(\cdot)) \\ &= \int_s^T \Big[\int_{[-\delta,0]^2} \Big(\langle Q(t,\theta,\theta')X(t+\theta), X(t+\theta')\rangle \\ &\qquad\qquad +2\langle S(t,\theta,\theta')X(t+\theta), u(t+\theta')\rangle \\ &\qquad\qquad +\langle R(t,\theta,\theta')u(t+\theta), u(t+\theta')\rangle \Big)\nu(d\theta)\nu(d\theta') \\ &\quad +2\int_{[-\delta,0]} \Big(\langle q(t,\theta), X(t+\theta)\rangle + \langle \rho(t,\theta), u(t+\theta)\rangle \Big)\nu(d\theta) \Big]dt \\ &\quad +\int_{[-\delta,0]^2} \langle G(\theta,\theta')X(T+\theta), X(T+\theta')\rangle \tilde{\nu}(d\theta)\tilde{\nu}(d\theta') \\ &\quad +2\int_{[-\delta,0]} \langle g(\theta), X(T+\theta)\rangle \tilde{\nu}(d\theta). \end{aligned} \tag{2.2.1}$$

Consider the case that (2.2.1) only contains the distributed delays, in other words, suppose $\nu(d\theta) = f_1(\theta)d\theta + f_2(\theta)\mu_{\{0\}}(d\theta)$ for some functions $f_1(\cdot), f_2(\cdot) \in L^\infty(-\delta, 0; \mathbb{R})$, where $\mu_{\{0\}}(d\theta)$ is the Dirac measure centered at 0, then without loss of generality, the every term of the cost functional (2.2.1) can be written separately as follows:

$$\begin{aligned}&\int_{[-\delta,0]^2}\langle Q(t,\theta,\theta')X(t+\theta), X(t+\theta')\rangle\nu(d\theta)\nu(d\theta')\\&:= \big\langle Q_{00}(t)X(t), X(t)\big\rangle + 2\int_{-\delta}^0\big\langle Q_{10}(t,\theta)^\top X(t+\theta), X(t)\big\rangle d\theta\\&\quad+\int_{[-\delta,0]^2}\big\langle Q_{11}(t,\theta,\theta')X(t+\theta), X(t+\theta')\big\rangle d\theta' d\theta,\end{aligned} \tag{2.2.2}$$

$$\begin{aligned}&\int_{[-\delta,0]^2}\langle S(t,\theta,\theta')X(t+\theta), u(t+\theta')\rangle\nu(d\theta)\nu(d\theta')\\&:=\big\langle S_{00}(t)X(t), u(t)\big\rangle+\int_{[-\delta,0]^2}\big\langle S_{11}(t,\theta,\theta')X(t+\theta), u(t+\theta')\big\rangle d\theta' d\theta\\&\quad+\int_{-\delta}^0\big\langle S_{01}(t,\theta)X(t+\theta), u(t)\big\rangle d\theta+\int_{-\delta}^0\big\langle S_{10}(t,\theta)^\top u(t+\theta), X(t)\big\rangle d\theta,\end{aligned} \tag{2.2.3}$$

$$\begin{aligned}&\int_{[-\delta,0]^2}\langle R(t,\theta,\theta')u(t+\theta), u(t+\theta')\rangle\nu(d\theta)\nu(d\theta')\\&:= \big\langle R_{00}(t)u(t), u(t)\big\rangle + 2\int_{-\delta}^0\big\langle R_{10}(t,\theta)^\top u(t+\theta), u(t)\big\rangle d\theta\\&\quad+\int_{[-\delta,0]^2}\big\langle R_{11}(t,\theta,\theta')u(t+\theta), u(t+\theta')\big\rangle d\theta' d\theta,\end{aligned} \tag{2.2.4}$$

$$\begin{aligned}&\int_{[-\delta,0]}\langle q(t,\theta), X(t+\theta)\rangle\nu(d\theta)\\&:= \big\langle q_0(t), X(t)\big\rangle + \int_{-\delta}^0\big\langle q_1(t,\theta), X(t+\theta)\big\rangle d\theta,\end{aligned} \tag{2.2.5}$$

$$\begin{aligned}&\int_{[-\delta,0]}\langle \rho(t,\theta), u(t+\theta)\rangle\nu(d\theta)\\&:= \big\langle \rho_0(t), u(t)\big\rangle + \int_{-\delta}^0\big\langle \rho_1(t,\theta), u(t+\theta)\big\rangle d\theta,\end{aligned} \tag{2.2.6}$$

$$\begin{aligned}&\int_{[-\delta,0]^2}\langle G(\theta,\theta')X(T+\theta), X(T+\theta')\rangle\tilde{\nu}(d\theta)\tilde{\nu}(d\theta')\\&:= \big\langle G_{00}X(T), X(T)\big\rangle + 2\int_{-\delta}^0\big\langle G_{10}(\theta)^\top X(T+\theta), X(T)\big\rangle d\theta\\&\quad+\int_{[-\delta,0]^2}\big\langle G_{11}(\theta,\theta')X(T+\theta), X(T+\theta')\big\rangle d\theta' d\theta,\end{aligned} \tag{2.2.7}$$

$$\int_{[-\delta,0]} \langle g(\theta), X(T+\theta)\rangle \tilde{\nu}(d\theta) \\ := \big\langle g_0, X(T)\big\rangle + \int_{-\delta}^{0} \big\langle g_1(\theta), X(T+\theta)\big\rangle d\theta, \tag{2.2.8}$$

where $Q_{00}(\cdot)$, $Q_{10}(\cdot\,,\cdot)$, $Q_{11}(\cdot\,,\cdot\,,\cdot)$, $S_{00}(\cdot)$, $S_{11}(\cdot\,,\cdot,\cdot)$, $S_{01}(\cdot\,,\cdot)$, $S_{10}(\cdot\,,\cdot)$, $R_{00}(\cdot)$, $R_{10}(\cdot\,,\cdot)$, $R_{11}(\cdot\,,\cdot\,,\cdot)$, $G_{10}(\cdot)$, $G_{11}(\cdot\,,\cdot)$ are the matrix-valued functions of proper dimensions, $q_0(\cdot)$, $q_1(\cdot,\cdot)$, $\rho_0(\cdot)$, $\rho_1(\cdot,\cdot)$, $g_1(\cdot)$ are vector-valued functions with appropriate dimensions, G_{00} is a symmetric matrix and $g_0 \in \mathbb{R}^n$.

Let us give the following hypothesis.

(H2.2) Let $T \geqslant \delta > 0$ be given.

(i) The coefficients of the state Eq. (2.1.1) satisfy the following assumptions:

$$A(\cdot\,,\cdot)\in C([0,T+\delta]\times[-\delta,0];\mathbb{R}^{n\times n}),$$
$$B(\cdot\,,\cdot)\in C([0,T+\delta]\times[-\delta,0];\mathbb{R}^{n\times m}), \quad b(\cdot)\in L^1(0,T;\mathbb{R}^n).$$

(ii) The weight coefficients of the cost functional (2.2.1) satisfy the following:

$$\begin{aligned}
&Q_{00}(\cdot)\in L^\infty(0,T;\mathbb{S}^n),\ Q_{10}(\cdot\,,\cdot)\in L^\infty([0,T]\times[-\delta,0];\mathbb{R}^{n\times n}),\\
&Q_{11}(\cdot\,,\cdot\,,\cdot)\in L^\infty([0,T]\times[-\delta,0]^2;\mathbb{R}^{n\times n}),\\
&S_{00}(\cdot)\in L^\infty(0,T;\mathbb{R}^{m\times n}),\ S_{01}(\cdot\,,\cdot), S_{10}(\cdot\,,\cdot)\in L^\infty([0,T]\times[-\delta,0];\mathbb{R}^{m\times n}),\\
&S_{11}(\cdot\,,\cdot\,,\cdot)\in L^\infty([0,T]\times[-\delta,0]^2;\mathbb{R}^{m\times n}),\\
&R_{00}(\cdot)\in L^\infty(0,T;\mathbb{S}^m),\ R_{10}(\cdot\,,\cdot)\in L^\infty([0,T]\times[-\delta,0];\mathbb{R}^{m\times m}),\\
&R_{11}(\cdot\,,\cdot,\cdot)\in L^\infty([0,T]\times[-\delta,0]^2;\mathbb{R}^{m\times m}),\\
&q_0(\cdot)\in L^1(0,T;\mathbb{R}^n),\ q_1(\cdot\,,\cdot)\in L^\infty([0,T]\times[-\delta,0];\mathbb{R}^n),\\
&\rho_0(\cdot)\in L^2(0,T;\mathbb{R}^m),\ \rho_1(\cdot\,,\cdot)\in L^\infty([0,T]\times[-\delta,0];\mathbb{R}^m),\\
&G_{00}\in\mathbb{S}^n,\ G_{10}(\cdot)\in L^2(-\delta,0;\mathbb{R}^{n\times n}),\\
&G_{11}(\cdot\,,\cdot)\in L^\infty([0,T]\times[-\delta,0];\mathbb{R}^{n\times n}),\\
&g_0\in\mathbb{R}^n,\ g_1(\cdot)\in L^2(-\delta,0;\mathbb{R}^n).
\end{aligned}$$

Moreover,

$$Q_{11}(t,\theta,\theta')^\top = Q_{11}(t,\theta',\theta),\ R_{11}(t,\theta,\theta')^\top = R_{11}(t,\theta',\theta),$$
$$G_{11}(\theta,\theta')^\top = G_{11}(\theta',\theta).$$

Under (H2.2), for any initial $(s,x,\varphi(\cdot),\psi(\cdot))\in[0,T)\times Z$ and any admissible control $u(\cdot)\in L^2(s,T;\mathbb{R}^m)$, by Theorem 2.1.1, the ODDE (2.1.1) admits a unique solution $X(\cdot)\equiv X(\cdot\,;s,x,\varphi(\cdot),\psi(\cdot),u(\cdot))\in C([s,T];\mathbb{R}^n)$. Therefore the cost functional (2.2.1) is well-defined. Now we rewrite Problem (P) in Sect. 1.2 as follows:

Problem (P) For any $(s, x, \varphi(\cdot), \psi(\cdot)) \in [0, T) \times Z$, to find a $u^*(\cdot) \in L^2(s, T; \mathbb{R}^m)$ such that (2.1.1) is satisfied and

$$
\begin{aligned}
& J(s, x, \varphi(\cdot), \psi(\cdot); u^*(\cdot)) \\
&= \inf_{u(\cdot)\in L^2(s,T;\mathbb{R}^m)} J(s, x, \varphi(\cdot), \psi(\cdot); u(\cdot)) := V(s, x, \varphi(\cdot), \psi(\cdot)),
\end{aligned}
$$

where $J(s, x, \varphi(\cdot), \psi(\cdot); u(\cdot))$ is defined by (2.2.1) and each of its terms is defined by (2.2.2)–(2.2.8). Any $u^*(\cdot) \in L^2(s, T; \mathbb{R}^m)$ that achieves the above infimum is called an *optimal open-loop control* for the initial data $(s, x, \varphi(\cdot), \psi(\cdot))$ and the corresponding solution $X^*(\cdot) \equiv X(\cdot\,; s, x, \varphi(\cdot), \psi(\cdot), u^*(\cdot))$ is called the *optimal state trajectory*. The function $V(\cdot\,,\cdot\,,\cdot)$ is called the *value function* of Problem (P). In the special case when $b(\cdot)$, $q_0(\cdot)$, $q_1(\cdot\,,\cdot)$, $\rho_0(\cdot)$, $\rho_1(\cdot\,,\cdot)$, g_0 and $g_1(\cdot)$ vanish, denote the corresponding delayed deterministic linear quadratic problem, cost functional, and the value function by Problem (P_0), $J_0(s, x, \varphi(\cdot), \psi(\cdot); u(\cdot))$ and $V_0(s, x, \varphi(\cdot), \psi(\cdot))$, respectively.

Now we would like to rewrite the cost functional (2.2.1) in terms of $\mathbf{Z}(\cdot)$ and $u(\cdot)$, before that we need to define some bounded linear operators. For simplicity, we still denote $L^2 := L^2(-\delta, 0; \mathbb{R}^k)$, $k = n, m$, which will be clear from the context.

$$
\begin{aligned}
&\tilde{Q}_{00}(t) : \mathbb{R}^n \to \mathbb{R}^n, && x \mapsto Q_{00}(t)x, \\
&\tilde{Q}_{01}(t) : L^2 \to \mathbb{R}^n, && \varphi(\cdot) \mapsto \int_{-\delta}^{0} Q_{10}(t,\theta)^\top \varphi(\theta)d\theta, \\
&\tilde{Q}_{10}(t) : \mathbb{R}^n \to L^2, && x \mapsto Q_{10}(t,\cdot)x, \\
&\tilde{Q}_{11}(t) : L^2 \to L^2, && \varphi(\cdot) \mapsto \int_{-\delta}^{0} Q_{11}(t,\theta,\cdot)\varphi(\theta)d\theta, \\
&\tilde{S}_{00}(t) : \mathbb{R}^n \to \mathbb{R}^m, && x \mapsto S_{00}(t)x, \\
&\tilde{S}_{01}(t) : L^2 \to \mathbb{R}^m, && \varphi(\cdot) \mapsto \int_{-\delta}^{0} S_{01}(t,\theta)\varphi(\theta)d\theta, \\
&\tilde{S}_{10}(t) : \mathbb{R}^n \to L^2, && x \mapsto S_{10}(t,\cdot)x, \\
&\tilde{S}_{11}(t) : L^2 \to L^2, && \varphi(\cdot) \mapsto \int_{-\delta}^{0} S_{11}(t,\theta,\cdot)\varphi(\theta)d\theta, \\
&\tilde{R}_{00}(t) : \mathbb{R}^m \to \mathbb{R}^m, && y \mapsto R_{00}(t)y, \\
&\tilde{R}_{01}(t) : L^2 \to \mathbb{R}^m, && \psi(\cdot) \mapsto \int_{-\delta}^{0} R_{10}(t,\theta)^\top \psi(\theta)d\theta, \\
&\tilde{R}_{10}(t) : \mathbb{R}^m \to L^2, && y \mapsto R_{10}(t,\cdot)y,
\end{aligned}
$$

$$
\begin{aligned}
&\tilde{R}_{11}(t): L^2 \to L^2, && \psi(\cdot) \mapsto \int_{-\delta}^{0} R_{11}(t,\theta,\cdot)\psi(\theta)d\theta,\\
&\tilde{G}_{00}: \mathbb{R}^n \to \mathbb{R}^n, && x \mapsto G_{00}x,\\
&\tilde{G}_{01}: L^2 \to \mathbb{R}^n, && \varphi(\cdot) \mapsto \int_{-\delta}^{0} G_{10}(\theta)^\top \varphi(\theta)d\theta,\\
&\tilde{G}_{10}: \mathbb{R}^n \to L^2, && x \mapsto G_{10}(\cdot)x,\\
&\tilde{G}_{11}: L^2 \to L^2, && \varphi(\cdot) \mapsto \int_{-\delta}^{0} G_{11}(\theta,\cdot)\varphi(\theta)d\theta.
\end{aligned}
$$

It is easy to verify that $\tilde{Q}_{01}(t)^* = \tilde{Q}_{10}(t)$, $\tilde{R}_{01}(t)^* = \tilde{R}_{10}(t)$, $\tilde{G}_{01}^* = \tilde{G}_{10}$, but $\tilde{S}_{01}(t)^* \neq \tilde{S}_{10}(t)$, in general. Denote

$$
\begin{aligned}
&\tilde{Q}(t) := \begin{bmatrix} \tilde{Q}_{00}(t) & \tilde{Q}_{01}(t) \\ \tilde{Q}_{10}(t) & \tilde{Q}_{11}(t) \end{bmatrix}, \qquad \tilde{S}(t) := \begin{bmatrix} \tilde{S}_{00}(t) & \tilde{S}_{01}(t) \\ \tilde{S}_{10}(t) & \tilde{S}_{11}(t) \end{bmatrix},\\
&\tilde{R}(t) := \begin{bmatrix} \tilde{R}_{00}(t) & \tilde{R}_{01}(t) \\ \tilde{R}_{10}(t) & \tilde{R}_{11}(t) \end{bmatrix}, \qquad \tilde{G} := \begin{bmatrix} \tilde{G}_{00} & \tilde{G}_{01} \\ \tilde{G}_{10} & \tilde{G}_{11} \end{bmatrix},\\
&\tilde{q}_0(t) := q_0(t), \qquad \tilde{q}_1(t) := q_1(t,\cdot), \qquad \tilde{\rho}_0(t) := \rho_0(t),\\
&\tilde{\rho}_1(t) := \rho_1(t,\cdot), \qquad \tilde{g}_0 := g_0, \qquad \tilde{g}_1 := g_1(\cdot),
\end{aligned}
$$

$$
\tilde{q}(t) := \begin{bmatrix} \tilde{q}_0(t) \\ \tilde{q}_1(t) \end{bmatrix}, \qquad \tilde{\rho}(t) := \begin{bmatrix} \tilde{\rho}_0(t) \\ \tilde{\rho}_1(t) \end{bmatrix}, \qquad \tilde{g} := \begin{bmatrix} \tilde{g}_0 \\ \tilde{g}_1 \end{bmatrix},
$$

then the cost functional (2.2.1) can be written as follows

$$
\begin{aligned}
&J(s,x,\varphi(\cdot),\psi(\cdot);u(\cdot))\\
&= \int_s^T \Big[\langle \tilde{Q}(t)\mathbf{X}(t), \mathbf{X}(t)\rangle + 2\Big\langle \tilde{S}(t)\mathbf{X}(t), \begin{pmatrix} u(t) \\ \mathbf{Y}_t \end{pmatrix} \Big\rangle\\
&\quad + \Big\langle \tilde{R}(t) \begin{pmatrix} u(t) \\ \mathbf{Y}_t \end{pmatrix}, \begin{pmatrix} u(t) \\ \mathbf{Y}_t \end{pmatrix} \Big\rangle + 2\langle \tilde{q}(t), \mathbf{X}(t)\rangle \\
&\quad + 2\Big\langle \tilde{\rho}(t), \begin{pmatrix} u(t) \\ \mathbf{Y}_t \end{pmatrix} \Big\rangle \Big] dt + \langle \tilde{G}\mathbf{X}(T), \mathbf{X}(T)\rangle + 2\langle \tilde{g}, \mathbf{X}(T)\rangle.
\end{aligned}
\tag{2.2.9}
$$

In the above, $\langle \cdot\,,\cdot\rangle$ has the different meaning. Next we define

$$
\begin{aligned}
&\tilde{S}_0(t): M^2 \longrightarrow \mathbb{R}^m, && (x^\top, \varphi^\top)^\top \mapsto \tilde{S}_{00}(t)x + \tilde{S}_{01}(t)\varphi,\\
&\tilde{S}_1(t): M^2 \longrightarrow L^2, && (x^\top, \varphi^\top)^\top \mapsto \tilde{S}_{10}(t)x + \tilde{S}_{11}(t)\varphi,
\end{aligned}
$$

$$
\mathbf{Q}(t) := \begin{bmatrix} \tilde{Q}(t) & \tilde{S}_1(t)^* \\ \tilde{S}_1(t) & \tilde{R}_{11}(t) \end{bmatrix} \in \mathscr{L}(Z), \quad \mathbf{S}(t) := \begin{bmatrix} \tilde{S}_0(t) & \tilde{R}_{01}(t) \end{bmatrix} \in \mathscr{L}(Z, \mathbb{R}^m),
$$

$$\mathbf{R}(t) := \tilde{R}_{00}(t) \in \mathbb{R}^{m\times m}, \qquad \mathbf{q}(t) := \begin{bmatrix} \tilde{q}(t) \\ \tilde{\rho}_1(t) \end{bmatrix} \in Z,$$

$$\boldsymbol{\rho}(t) := \tilde{\rho}_0(t) \in \mathbb{R}^m, \qquad \mathbf{G} := \begin{bmatrix} \tilde{G} & 0 \\ 0 & 0 \end{bmatrix} \in \mathscr{L}(Z), \qquad \mathbf{g} := \begin{bmatrix} \tilde{g} \\ 0 \end{bmatrix} \in Z.$$

Then (2.2.9) can further be written

$$\begin{aligned} J(s, \mathbf{Z}_0; u(\cdot)) = \int_s^T &\Big[\big\langle \mathbf{Q}(t)\mathbf{Z}(t), \mathbf{Z}(t)\big\rangle_Z + 2\big\langle \mathbf{S}(t)\mathbf{Z}(t), u(t)\big\rangle \\ &+\big\langle \mathbf{R}(t)u(t), u(t)\big\rangle + 2\big\langle \mathbf{q}(t), \mathbf{Z}(t)\big\rangle_Z + 2\big\langle \boldsymbol{\rho}(t), u(t)\big\rangle\Big]dt \\ &+\big\langle \mathbf{G}\mathbf{Z}(T), \mathbf{Z}(T)\big\rangle_Z + 2\big\langle \mathbf{g}, \mathbf{Z}(T)\big\rangle_Z. \end{aligned} \tag{2.2.10}$$

Noting $\mathbf{X}(\cdot) \in C([s, T]; M^2)$, $Y_\cdot \in C([s, T]; L^2)$ and $\mathbf{Z}(\cdot) \in C([s, T]; Z)$, under (H2.2) the cost functional (2.2.9) and (2.2.10) are meaningful. Thus, we have transformed Problem (P) into the LQ problem associated with (2.1.29) (or (2.1.27)) and (2.2.10). More precisely, we formulate the problem as follows.

Problem (EP) For any $(s, \mathbf{Z}_0) \in [0, T) \times Z$, find a $u^*(\cdot) \in L^2(s, T; \mathbb{R}^m)$ such that (2.1.29) (or (2.1.27)) is satisfied and

$$J(s, \mathbf{Z}_0; u^*(\cdot)) = \inf_{u(\cdot)\in L^2(s,T;\mathbb{R}^m)} J(s, \mathbf{Z}_0; u(\cdot)) := V(s, \mathbf{Z}_0). \tag{2.2.11}$$

Similarly, any $u^*(\cdot) \in L^2(s, T; \mathbb{R}^m)$ that achieves the above infimum is called an *optimal open-loop control* for the initial pair $(s, \mathbf{Z}_0)$, and the corresponding solution $\mathbf{Z}^*(\cdot) \equiv \mathbf{Z}(\cdot\,; s, \mathbf{Z}_0, u^*(\cdot))$ is called the *optimal state*. $V(\cdot\,, \cdot)$ is called the *value function* of Problem (EP). In the special case when $b(\cdot)$, $\mathbf{q}(\cdot)$, $\boldsymbol{\rho}(\cdot)$ and $\mathbf{g}$ vanish, we denote the corresponding LQ problem, the cost functional and the value function by Problem (EP_0), $J_0(s, \mathbf{Z}_0; u(\cdot))$ and $V_0(s, \mathbf{Z}_0)$, respectively.

Remark 2.2.1 By (2.1.21), (2.1.22), (2.1.26), Lemma 2.1.4 and Theorem 2.1.2, Problem (P) is equivalent to Problem (EP).

Remark 2.2.2 Considering lifted models of delayed linear quadratic optimal control problems, our model can cover most cases in the existing literature, such as [1], [2], [3], [4], [5], [6], [7], [8] (time-varying coefficients, state pointwise delays and state distributed delays), [11] (time-invariant coefficients, state pointwise delays and state distributed delays), [13] (time-invariant coefficients, control pointwise delays and control distributed delays), [17] (time-invariant coefficients, state pointwise delays, state distributed delays and control point delays).

Next we give the following proposition for $\mathbf{T}(t, s)$ (see (2.1.28) for its definition), which plays an important role in the study of Riccati equations in Sects. 3.2 and 3.3.

Proposition 2.2.3 *Let* (H2.2)(i) *hold. Then* $\mathbf{T}(\cdot\,,\cdot)^*$ *is strongly continuous.*

Proof

Step 1: we show that $[\Phi(t,s)^*\xi]^0$ is continuous in t and s, for any given $\xi \in M^2$.

Recall (2.1.11), $\mathfrak{X}(\cdot) \equiv \mathfrak{X}(\cdot\,; s, x, \varphi(\cdot))$ is the solution to (2.1.1) with $B(\cdot\,,\cdot) = 0$ and $b(\cdot) = 0$. And by the variation of constants formula (2.1.4), we have

$$\begin{aligned}\mathfrak{X}(t) &= \Phi^0(t,s)x + \int_s^{t\wedge(s+\delta)} \Phi^0(t,r)\int_{[-\delta,s-r)} A(r,\theta)\varphi(r+\theta-s)\mu(d\theta)dr\\ &= \Phi^0(t,s)x + \int_{-\delta}^0 \Phi^1(t,s,\theta)\varphi(\theta)d\theta,\end{aligned}$$

where

$$\Phi^0(t,s)x := X_0(t;s,x) = \mathfrak{X}(t;s,x,0), \tag{2.2.12}$$

$$\Phi^1(t,s,\theta)x := \int_{[-\delta,\theta]} \Phi^0(t,s+\theta-\tau)A(s+\theta-\tau,\tau)x\mathbf{1}_{(s+\theta-t,0]}(\tau)\mu(d\tau). \tag{2.2.13}$$

Then we get

$$\Phi(t,s)^*\xi := \begin{pmatrix}[\Phi(t,s)^*\xi]^0\\ [\Phi(t,s)^*\xi]^1\end{pmatrix}, \quad \forall\, \xi = \begin{pmatrix}x\\ \varphi\end{pmatrix} \in M^2,$$

where

$$\begin{aligned}[\Phi(t,s)^*\xi]^0 &= \Phi^0(t,s)^*x + \int_{(-\delta)\vee(s-t)}^0 \Phi^0(t+\theta,s)^*\varphi(\theta)d\theta,\\ [\Phi(t,s)^*\xi]^1(\theta) &= \Phi^1(t,s,\theta)^*x + \int_{(-\delta)\vee(s-t)}^0 \Phi^1(t+\beta,s,\theta)^*\varphi(\beta)d\beta\\ &\quad +\varphi(s+\theta-t)\mathbf{1}_{(t-s-\delta,0]}(\theta).\end{aligned} \tag{2.2.14}$$

Here

$$\Phi^1(t,s,\theta)^*x = \int_{[-\delta,\theta]} A(s+\theta-\tau,\tau)^\top\Phi^0(t,s+\theta-\tau)^*x\mathbf{1}_{(s+\theta-t,0]}(\tau)\mu(d\tau). \tag{2.2.15}$$

By (2.2.12), we claim that $\Phi^0(t,s)^*x = p(s;t,x,0)$, where $p(\cdot\,;t,x,0)$ is the solution to the following equation:

$$\begin{cases} \dot p(r) + \int_{[-\delta,0]} A(r-\theta,\theta)^\top p(r-\theta)\mu(d\theta) = 0, \quad r\in[s,t],\\ p(t)=x,\ p(r)=0, r\in(t,t+\delta]. \end{cases}$$

In fact, for any $x,y\in\mathbb{R}^n$, noting $\int_s^t \frac{d}{dr}\langle \mathfrak{X}(r;s,x,0), p(r;t,y,0)\rangle = 0$, thus we obtain $\langle \mathfrak{X}(t;s,x,0),y\rangle = \langle x, p(s;t,y,0)\rangle$. Since $p(s;t,y,0)$ is continuous in t and s, $\Phi^0(\cdot,\cdot)^*$ is strongly continuous and thus $[\Phi(t,s)^*\xi]^0$ is continuous in t and s. Moreover, noting $A(\cdot,\cdot)$ is continuous, so (2.2.13) and (2.2.15) are well-defined.

Step 2: we show that $[\Phi(t,s)^*\xi]^1$ is continuous in t and s.

Take $0\leqslant s_1<s_2\leqslant t$, without loss of generality, assume $-\delta<s_1-t<s_2-t$, then we have

$$\begin{aligned}
&\int_{-\delta}^0 \left|\Phi^1(t,s_1,\theta)^*x - \Phi^1(t,s_2,\theta)^*x\right|^2 d\theta\\
&= \int_{-\delta}^0 \bigg|\int_{[-\delta,\theta]} \Big[A(s_1+\theta-\alpha,\alpha)^\top \Phi^0(t,s_1+\theta-\alpha)^*x\mathbf{1}_{(s_1+\theta-t,0]}(\alpha)\\
&\qquad -A(s_2+\theta-\alpha,\alpha)^\top\Phi^0(t,s_2+\theta-\alpha)^*x\mathbf{1}_{(s_2+\theta-t,0]}(\alpha)\Big]\mu(d\alpha)\bigg|^2 d\theta\\
&\leqslant K\int_{-\delta}^{t-\delta-s_2}\int_{[-\delta,\theta]}\Big|A(s_1+\theta-\alpha,\alpha)^\top\Phi^0(t,s_1+\theta-\alpha)^*x\\
&\qquad -A(s_2+\theta-\alpha,\alpha)^\top\Phi^0(t,s_2+\theta-\alpha)^*x\Big|^2\mu(d\alpha)d\theta\\
&\quad +K\int_{t-\delta-s_2}^{t-\delta-s_1}\int_{[-\delta,\theta]}\Big|A(s_1+\theta-\alpha,\alpha)^\top\Phi^0(t,s_1+\theta-\alpha)^*x\\
&\qquad -A(s_2+\theta-\alpha,\alpha)^\top\Phi^0(t,s_2+\theta-\alpha)^*x\mathbf{1}_{(s_2+\theta-t,0]}(\alpha)\Big|^2\mu(d\alpha)d\theta\\
&\quad +K\int_{t-\delta-s_1}^0\int_{(s_1+\theta-t,s_2+\theta-t]}|A(s_1+\theta-\alpha,\alpha)^\top\Phi^0(t,s_1+\theta-\alpha)^*x|^2\mu(d\alpha)d\theta\\
&\quad +K\int_{t-\delta-s_1}^0\int_{(s_2+\theta-t,\theta]}\Big|A(s_1+\theta-\alpha,\alpha)^\top\Phi^0(t,s_1+\theta-\alpha)^*x\\
&\qquad -A(s_2+\theta-\alpha,\alpha)^\top\Phi^0(t,s_2+\theta-\alpha)^*x\Big|^2\mu(d\alpha)d\theta.
\end{aligned}$$

Recall that $K>0$ is a generic constant. By the continuity of $A(\cdot\,,\cdot)$ and $\Phi^0(\cdot\,,\cdot)^*x$, $\int_{-\delta}^0|\Phi^1(t,s_1,\theta)^*x - \Phi^1(t,s_2,\theta)^*x|^2d\theta$ can be made arbitrarily small by choosing a small enough distance $|s_1-s_2|$. Similarly, noting $A(\cdot\,,\cdot)$ and $\Phi^0(\cdot\,,\cdot)^*$ are bounded, $\varphi(\cdot)\in L^2$ and $\Phi^0(\cdot\,,\cdot)^*$ is strongly continuous, we have

$$\int_{-\delta}^0\bigg|\int_{(-\delta)\vee(s_1-t)}^0 \Phi^1(t+\alpha,s_1,\theta)^*\varphi(\alpha)d\alpha - \int_{(-\delta)\vee(s_2-t)}^0 \Phi^1(t+\alpha,s_2,\theta)^*\varphi(\alpha)d\alpha\bigg|^2 d\theta$$

and

$$\int_{-\delta}^{0} \left|\varphi(s_1+\theta-t)\mathbf{1}_{(t-s_1-\delta,0]}(\theta) - \varphi(s_2+\theta-t)\mathbf{1}_{(t-s_2-\delta,0]}(\theta)\right|^2 d\theta$$

can both be made arbitrarily small by choosing a small enough distance $|s_1 - s_2|$, thus $[\Phi(t,s)^*\xi]^1$ is continuous in s on $[0,t]$ and then similarly we can show that $[\Phi(t,s)^*\xi]^1$ is continuous in t on $[s,T]$.

Step 3: we show that $\mathbf{T}(\cdot\,,\cdot)^*$ is strongly continuous.

It is easy to verify that

$$\mathbf{T}(t,s)^*z = \begin{pmatrix}[\mathbf{T}(t,s)^*z]^0\\ [\mathbf{T}(t,s)^*z]^1\end{pmatrix}, \quad \forall\, z = \begin{pmatrix}\xi\\ \psi\end{pmatrix} \in Z, \tag{2.2.16}$$

where

$$\begin{aligned}&[\mathbf{T}(t,s)^*z]^0 = \Phi(t,s)^*\xi,\\ &[\mathbf{T}(t,s)^*z]^1 = e^{D^*(t-s)}\psi + \int_{[s+\cdot-t,\cdot]} B(s+\cdot-\alpha,\alpha)^\top[\Phi(t,s+\cdot-\alpha)^*\xi]^0\mu(d\alpha).\end{aligned}$$

and e^{D^*t} is the adjoint operator of e^{Dt}. By Step 1-2, $\Phi(\cdot,\cdot)^*$ is strongly continuous. Since $B(\cdot\,,\cdot)$ is continuous, $[\mathbf{T}(t,s)^*z]^1$ is well-defined. Noting $B(\cdot,\cdot)$ and $\Phi(\cdot\,,\cdot)^*$ are bounded, $\Phi(\cdot\,,\cdot)^*$ and $e^{D^*\cdot}$ are strongly continuous, by (2.1.28) we conclude that $\mathbf{T}(\cdot\,,\cdot)^*$ is strongly continuous. This completes the proof.

□

Next we introduce the following definitions.

Definition 2.2.4 Problem (P) is said to be

(i) *(uniquely) open-loop solvable at initial data* $(s,x,\varphi(\cdot),\psi(\cdot)) \in [0,T]\times Z$, if there exists a (unique) $u^*(\cdot) \in L^2(s,T;\mathbb{R}^m)$ satisfying (2.2.11).
(ii) *(uniquely) open-loop solvable at some* $s \in [0,T)$, if for any $(x,\varphi(\cdot),\psi(\cdot)) \in Z$, there exists a (unique) $u^*(\cdot) \in L^2(s,T;\mathbb{R}^m)$ satisfying (2.2.11).
(iii) *(uniquely) open-loop solvable on* $[s,T)$, if it is (uniquely) open-loop solvable at all $t \in [s,T)$.

Next, we study the unique solvability of the following equation:

$$\mathbf{Z}(t) = \mathbf{T}(t,s)\mathbf{Z}_0 + \int_s^t \mathbf{T}(t,r)\Big\{\mathbf{B}\big[\Theta(r)\mathbf{Z}(r) + v(r)\big] + \mathbf{b}(r)\Big\}dr, \tag{2.2.17}$$

noting that $\mathbf{B} \notin \mathscr{L}(\mathbb{R}^m, Z)$. First we present the following lemma, which plays an important role in the study of unique solvability for (2.2.17).

Lemma 2.2.5 *Let* (H2.2)(i) *hold,* $\Theta(\cdot) : [0, T] \to \mathscr{L}(Z, \mathbb{R}^m)$ *is strongly continuous (i.e.,* $\Theta(\cdot)z$ *is continuous on* $[0, T]$ *for all* $z \in Z$*). Suppose* $\sup_{0\leqslant t\leqslant T} \|\Theta(t)\|_{\mathscr{L}(Z,\mathbb{R}^m)} < \infty$ *and* $\mathbf{Z}_0 \in Z$*, then the integral equation*

$$\mathbf{T}_\Theta(t, s)\mathbf{Z}_0 = \mathbf{T}(t, s)\mathbf{Z}_0 + \int_s^t \mathbf{T}(t, r)\mathbf{B}\Theta(r)\mathbf{T}_\Theta(r, s)\mathbf{Z}_0 dr,\ 0\leqslant s\leqslant t\leqslant T,$$

has a unique solution with the following properties:

(i) $\mathbf{T}_\Theta(\cdot, \cdot) : \Delta_*[0, T] \to \mathscr{L}(Z)$ *is a mild evolution operator.*
(ii) *Define*

$$\begin{cases} \mathbf{T}'_0(t, s)\mathbf{Z}_0 := \mathbf{T}(t, s)\mathbf{Z}_0, \\ \mathbf{T}'_k(t, s)\mathbf{Z}_0 := \displaystyle\int_s^t \mathbf{T}'_{k-1}(t, r)\mathbf{B}\Theta(r)\mathbf{T}(r, s)\mathbf{Z}_0 dr, \quad k = 1, 2, \cdots, \end{cases}$$

which is well-defined. Moreover, define

$$\int_s^t \mathbf{T}_\Theta(t, r)\mathbf{B}u dr := \sum_{k=0}^{\infty} \int_s^t \mathbf{T}'_k(t, r)\mathbf{B}u dr, \quad \forall\, u \in \mathbb{R}^m, \tag{2.2.18}$$

then

$$\mathbf{T}_\Theta(t, s)\mathbf{Z}_0 = \mathbf{T}(t,s)\mathbf{Z}_0 + \int_s^t \mathbf{T}_\Theta(t,r)\mathbf{B}\Theta(r)\mathbf{T}(r,s)\mathbf{Z}_0 dr,\ 0\leqslant s\leqslant t\leqslant T. \tag{2.2.19}$$

Proof For simplicity of notation, we use $\|\cdot\|$ for different norms which can be identified from the context, such as $\|\cdot\|_{M^2}$, $\|\cdot\|_{L^2}$, etc. Suppose

$$\sup_{0\leqslant t\leqslant T} \|\Theta(t)\| + \sup_{0\leqslant s\leqslant t\leqslant T} \|\Phi^0(t, s)\| + \sup_{0\leqslant s\leqslant t\leqslant T} \|\mathbf{T}(t, s)\| + \sup_{\substack{0\leqslant t\leqslant T \\ -\delta\leqslant\theta\leqslant 0}} |B(t, \theta)| \leqslant K,$$

where $\Phi^0(t, s) \in \mathscr{L}(\mathbb{R}^n)$ is a bounded linear operator defined as (2.1.5). We split the proof into several steps.

Step 1. Existence. Define

$$\begin{cases} \mathbf{T}_0(t, s)\mathbf{Z}_0 := \mathbf{T}(t, s)\mathbf{Z}_0, \\ \mathbf{T}_k(t, s)\mathbf{Z}_0 := \displaystyle\int_s^t \mathbf{T}(t, r)\mathbf{B}\Theta(r)\mathbf{T}_{k-1}(r, s)\mathbf{Z}_0 dr, \quad k = 1, 2, \cdots. \end{cases}$$

Next we aim to show that

$$\mathbf{T}_\Theta(t,s)\mathbf{Z}_0 = \sum_{k=0}^{\infty} \mathbf{T}_k(t,s)\mathbf{Z}_0.$$

Denote $h(r,s) := \Theta(r)\mathbf{T}_{k-1}(r,s)\mathbf{Z}_0$. Noting

$$\begin{aligned} &\Big\| \int_s^t e^{D(t-r)}\mathcal{D}h(r,s)dr \Big\|^2 = \int_{(-\delta)\vee(s-t)}^0 |h(t+\theta,s)|^2 d\theta \\ &\leqslant K^2 \int_{(-\delta)\vee(s-t)}^0 \|\mathbf{T}_{k-1}(t+\theta,s)\mathbf{Z}_0\|^2 d\theta, \end{aligned} \tag{2.2.20}$$

and recalling the definition of $\widetilde{B}(\cdot)$ (see (2.1.13))

$$\begin{aligned} &\Big\| \int_s^t \int_r^t \Phi(t,\beta)\widetilde{B}(\beta)e^{D(\beta-r)}\mathcal{D}h(r,s)d\beta dr \Big\|^2 \\ &= \Big\| \int_s^t \Phi(t,r)\widetilde{B}(r) \int_s^r e^{D(r-\beta)}\mathcal{D}h(\beta,s)d\beta dr \Big\|^2 \\ &= \Big\| \int_s^t \Phi(t,r) \begin{pmatrix} \int_{[-\delta,0]} B(r,\theta)h(r+\theta,s)\mathbf{1}_{(s-r,0]}(\theta)\mu(d\theta) \\ 0 \end{pmatrix} dr \Big\|^2 \\ &= \Big\| \int_{[-\delta,0]} \int_{s+\theta}^{t+\theta} \Phi^0(t,r-\theta)B(r-\theta,\theta)h(r,s)\mathbf{1}_{(s-r+\theta,0]}(\theta)dr\mu(d\theta) \Big\|^2 \\ &+ \Big\| \int_{[-\delta,0]} \int_{s+\theta}^{t+\theta} \Phi^0(t+\cdot,r-\theta)B(r-\theta,\theta)h(r,s)\mathbf{1}_{(s-r+\theta,0]}(\theta)dr\mu(d\theta) \Big\|^2 \\ &\leqslant K' \int_{[-\delta,0]} \int_{s-\theta}^t |h(r+\theta,s)|^2 dr\mu(d\theta) \leqslant K' \int_s^t \|\mathbf{T}_{k-1}(r,s)\mathbf{Z}_0\|^2 dr, \end{aligned}$$

hereafter K' is a generic constant, which and (2.2.20) yield

$$\begin{aligned} &\|\mathbf{T}_k(t,s)\mathbf{Z}_0\|^2 \leqslant K' \int_{(-\delta)\vee(s-t)}^0 \|\mathbf{T}_{k-1}(t+\theta,s)\mathbf{Z}_0\|^2 d\theta + K' \int_s^t \|\mathbf{T}_{k-1}(r,s)\mathbf{Z}_0\|^2 dr \\ &\leqslant K_0 \int_s^t \|\mathbf{T}_{k-1}(r,s)\mathbf{Z}_0\|^2 dr \leqslant K_0^2 \int_s^t \int_s^r \|\mathbf{T}_{k-2}(\beta,s)\mathbf{Z}_0\|^2 d\beta dr \\ &\leqslant K \frac{K_0^k (t-s)^k}{k!} \|\mathbf{Z}_0\|^2, \end{aligned}$$

where $K_0 > 0$ is an absolute constant independent of $k \geqslant 1$. Thus we deduce

$$\|\mathbf{T}_k(t,s)\mathbf{Z}_0\| \leqslant K \left(\frac{K_0^k (t-s)^k}{k!} \right)^{\frac{1}{2}} \|\mathbf{Z}_0\|.$$

It follows that

$$\|\sum_{k=0}^{\infty}\mathbf{T}_k(t,s)\mathbf{Z}_0\| \leqslant \sum_{k=0}^{\infty}\Big(\frac{K_0^k(t-s)^k}{k!}\Big)^{\frac{1}{2}} K\|\mathbf{Z}_0\| < \infty.$$

And apparently we get

$$\sup_{0\leqslant s\leqslant t\leqslant T} \|\sum_{k=0}^{\infty}\mathbf{T}_k(t,s)\mathbf{Z}_0\| < \infty.$$

Using similar approximating method (see (3.2.28)) in Sect. 3.2 and noting

$$\int_s^t \mathbf{T}(t,r)\mathbf{B}^{\lambda}u(r)dr \to \int_s^t \mathbf{T}(t,r)\mathbf{B}u(r)dr, \quad \text{as } \lambda\to\infty,$$

by Proposition 1.3.6 we derive

$$\begin{aligned}
\sum_{k=0}^{\infty}\mathbf{T}_k(t,s)\mathbf{Z}_0 &= \mathbf{T}(t,s)\mathbf{Z}_0 + \sum_{k=1}^{\infty}\int_s^t \mathbf{T}(t,r)\mathbf{B}\Theta(r)\mathbf{T}_{k-1}(r,s)\mathbf{Z}_0 dr\\
&= \mathbf{T}(t,s)\mathbf{Z}_0 + \sum_{k=1}^{\infty}\lim_{\lambda\to\infty}\int_s^t \mathbf{T}(t,r)\mathbf{B}^{\lambda}\Theta(r)\mathbf{T}_{k-1}(r,s)\mathbf{Z}_0 dr\\
&= \mathbf{T}(t,s)\mathbf{Z}_0 + \lim_{\lambda\to\infty}\sum_{k=1}^{\infty}\int_s^t \mathbf{T}(t,r)\mathbf{B}^{\lambda}\Theta(r)\mathbf{T}_{k-1}(r,s)\mathbf{Z}_0 dr\\
&= \mathbf{T}(t,s)\mathbf{Z}_0 + \lim_{\lambda\to\infty}\int_s^t \mathbf{T}(t,r)\mathbf{B}^{\lambda}\Theta(r)\sum_{k=1}^{\infty}\mathbf{T}_{k-1}(r,s)\mathbf{Z}_0 dr\\
&= \mathbf{T}(t,s)\mathbf{Z}_0 + \int_s^t \mathbf{T}(t,r)\mathbf{B}\Theta(r)\sum_{k=0}^{\infty}\mathbf{T}_k(r,s)\mathbf{Z}_0 dr,
\end{aligned}$$

which yields

$$\mathbf{T}_{\Theta}(t,s)\mathbf{Z}_0 = \sum_{k=0}^{\infty}\mathbf{T}_k(t,s)\mathbf{Z}_0.$$

It is apparent that $\mathbf{T}_{\Theta}(t,t)\mathbf{Z}_0 = \mathbf{T}(t,t)\mathbf{Z}_0 = \mathbf{Z}_0$ and $\sup_{0\leqslant s\leqslant t\leqslant T}\|\mathbf{T}_{\Theta}(t,s)\| < \infty$.

Step 2. Uniqueness. Suppose there is another solution $\mathbf{T}'_{\Theta}(t,s)$ and denote $\hat{\mathbf{T}}_{\Theta}(t,s) := \mathbf{T}_{\Theta}(t,s) - \mathbf{T}'_{\Theta}(t,s)$, then we have

$$\hat{\mathbf{T}}_{\Theta}(t,s)\mathbf{Z}_0 = \int_s^t \mathbf{T}(t,r)\mathbf{B}\Theta(r)\hat{\mathbf{T}}_{\Theta}(r,s)\mathbf{Z}_0 dr.$$

Repeating the steps in Step 1, we obtain

$$\|\hat{\mathbf{T}}_\Theta(t,s)\mathbf{Z}_0\|^2 \leqslant K_0 \int_s^t \|\hat{\mathbf{T}}_\Theta(r,s)\mathbf{Z}_0\|^2 dr,$$

which yields

$$\hat{\mathbf{T}}_\Theta(t,s)\mathbf{Z}_0 = 0, \text{ for any } \mathbf{Z}_0 \in Z.$$

Step 3. Semigroup property. Since

$$\begin{aligned}
&\mathbf{T}_\Theta(t,r)\mathbf{T}_\Theta(r,s)\mathbf{Z}_0 \\
&= \mathbf{T}(t,r)\mathbf{T}_\Theta(r,s)\mathbf{Z}_0 + \int_r^t \mathbf{T}(t,\beta)\mathbf{B}\Theta(\beta)\mathbf{T}_\Theta(\beta,r)\mathbf{T}_\Theta(r,s)\mathbf{Z}_0 d\beta \\
&= \mathbf{T}(t,r)\mathbf{T}(r,s)\mathbf{Z}_0 + \int_s^r \mathbf{T}(t,\beta)\mathbf{B}\Theta(\beta)\mathbf{T}_\Theta(\beta,s)\mathbf{Z}_0 d\beta \\
&\quad + \int_r^t \mathbf{T}(t,\beta)\mathbf{B}\Theta(\beta)\mathbf{T}_\Theta(\beta,r)\mathbf{T}_\Theta(r,s)\mathbf{Z}_0 d\beta,
\end{aligned}$$

it follows that

$$\begin{aligned}
&\mathbf{T}_\Theta(t,r)\mathbf{T}_\Theta(r,s)\mathbf{Z}_0 - \mathbf{T}_\Theta(t,s)\mathbf{Z}_0 \\
&= \int_r^t \mathbf{T}(t,\beta)\mathbf{B}\Theta(\beta)\big[\mathbf{T}_\Theta(\beta,r)\mathbf{T}_\Theta(r,s)\mathbf{Z}_0 - \mathbf{T}_\Theta(\beta,s)\mathbf{Z}_0\big] d\beta.
\end{aligned}$$

Repeating the steps in Step 1, we get

$$\begin{aligned}
&\|\mathbf{T}_\Theta(t,r)\mathbf{T}_\Theta(r,s)\mathbf{Z}_0 - \mathbf{T}_\Theta(t,s)\mathbf{Z}_0\|^2 \\
&\leqslant K_0 \int_r^t \|\mathbf{T}_\Theta(\beta,r)\mathbf{T}_\Theta(r,s)\mathbf{Z}_0 - \mathbf{T}_\Theta(\beta,s)\mathbf{Z}_0\|^2 d\beta,
\end{aligned}$$

which yields

$$\mathbf{T}_\Theta(t,r)\mathbf{T}_\Theta(r,s)\mathbf{Z}_0 = \mathbf{T}_\Theta(t,s)\mathbf{Z}_0.$$

Step 4. Continuity. Consider

$$\phi(t,s)\mathbf{Z}_0 = \int_s^t \mathbf{T}(t,r)\mathbf{B}\Theta(r)\mathbf{T}_\Theta(r,s)\mathbf{Z}_0 dr.$$

Take $h > 0$, $s_1 \in [s, s+\delta)$, noting that $\Theta(\cdot)$ is bounded, we have

$$\begin{aligned}
&\Big\| \int_s^{s_1+h} e^{D(s_1+h-r)} \mathcal{D}\Theta(r)\mathbf{T}_\Theta(r,s)\mathbf{Z}_0 dr \\
&\quad - \int_s^{s_1} e^{D(s_1-r)} \mathcal{D}\Theta(r)\mathbf{T}_\Theta(r,s)\mathbf{Z}_0 dr \Big\|^2 \\
&= \int_{-\delta}^{0} \Big| \Theta(s_1+h+\theta)\mathbf{T}_\Theta(s_1+h+\theta,s)\mathbf{Z}_0 \mathbf{1}_{[s-s_1-h,0]}(\theta) \\
&\qquad - \Theta(s_1+\theta)\mathbf{T}_\Theta(s_1+\theta,s)\mathbf{Z}_0 \mathbf{1}_{[s-s_1,0]}(\theta) \Big|^2 d\theta \\
&\leqslant 3K' \int_{-\delta}^{0} \Big\| \big[\mathbf{T}_\Theta(s_1+h+\theta,s) - \mathbf{T}_\Theta(s_1+\theta,s)\big]\mathbf{Z}_0 \mathbf{1}_{[s-s_1,0]}(\theta) \Big\|^2 d\theta \\
&\quad + 3\int_{-\delta}^{0} \Big| \big[\Theta(s_1+h+\theta) - \Theta(s_1+\theta)\big]\mathbf{T}_\Theta(s_1+\theta,s)\mathbf{Z}_0 \mathbf{1}_{[s-s_1,0]}(\theta) \Big|^2 d\theta \\
&\quad + 3\int_{s-s_1-h}^{s-s_1} \Big| \Theta(s_1+h+\theta)\mathbf{T}_\Theta(s_1+h+\theta,s)\mathbf{Z}_0 \Big|^2 d\theta.
\end{aligned} \tag{2.2.21}$$

In the following K' is a generic constant. Since $\Theta(\cdot)$ is strongly continuous and $\Theta(\cdot)$, $\mathbf{T}_\Theta(\cdot,\cdot)$ are bounded,

$$\begin{aligned}
&\int_{-\delta}^{0} \Big| \big[\Theta(s_1+h+\theta) - \Theta(s_1+\theta)\big]\mathbf{T}_\Theta(s_1+\theta,s)\mathbf{Z}_0 \mathbf{1}_{[s-s_1,0]}(\theta) \Big|^2 d\theta \to 0, \\
&\int_{s-s_1-h}^{s-s_1} |\Theta(s_1+h+\theta)\mathbf{T}_\Theta(s_1+h+\theta,s)\mathbf{Z}_0|^2 d\theta \to 0, \quad \text{as } h \to 0.
\end{aligned} \tag{2.2.22}$$

By the boundedness of $\Phi(\cdot,\cdot)$, $\Theta(\cdot)$, $\mathbf{T}_\Theta(\cdot,\cdot)$ and the strong continuity of $\Phi(\cdot,\cdot)$, we have

$$\begin{aligned}
&\Big\| \int_s^{s_1+h} \int_r^{s_1+h} \Phi(s_1+h,\beta)\widetilde{B}(\beta) e^{D(\beta-r)} \mathcal{D}\Theta(r)\mathbf{T}_\Theta(r,s)\mathbf{Z}_0 d\beta dr \\
&\quad - \int_s^{s_1} \int_r^{s_1} \Phi(s_1,\beta)\widetilde{B}(\beta) e^{D(\beta-r)} \mathcal{D}\Theta(r)\mathbf{T}_\Theta(r,s)\mathbf{Z}_0 d\beta dr \Big\|^2 \\
&= \Big\| \int_s^{s_1+h} \Phi(s_1+h,\beta)\widetilde{B}(\beta) \int_s^{\beta} e^{D(\beta-r)} \mathcal{D}\Theta(r)\mathbf{T}_\Theta(r,s)\mathbf{Z}_0 dr d\beta \\
&\quad - \int_s^{s_1} \Phi(s_1,\beta)\widetilde{B}(\beta) \int_s^{\beta} e^{D(\beta-r)} \mathcal{D}\Theta(r)\mathbf{T}_\Theta(r,s)\mathbf{Z}_0 dr d\beta \Big\|^2 \\
&\leqslant 2\Big\| \int_{s_1}^{s_1+h} \Phi(s_1+h,\beta)\widetilde{B}(\beta) \int_s^{\beta} e^{D(\beta-r)} \mathcal{D}\Theta(r)\mathbf{T}_\Theta(r,s)\mathbf{Z}_0 dr d\beta \Big\|^2 \\
&\quad + 2\Big\| \int_s^{s_1} \big[\Phi(s_1+h,\beta) - \Phi(s_1,\beta)\big]\widetilde{B}(\beta) \\
&\qquad \times \int_s^{\beta} e^{D(\beta-r)} \mathcal{D}\Theta(r)\mathbf{T}_\Theta(r,s)\mathbf{Z}_0 dr d\beta \Big\|^2 \to 0, \quad \text{as } h \to 0.
\end{aligned} \tag{2.2.23}$$

From (2.2.21)–(2.2.23), we deduce

$$\begin{aligned}
&\|\mathbf{T}_\Theta(s_1+h,s)\mathbf{Z}_0-\mathbf{T}_\Theta(s_1,s)\mathbf{Z}_0\|^2\\
&\leqslant 2\|\mathbf{T}(s_1+h,s)\mathbf{Z}_0-\mathbf{T}(s_1,s)\mathbf{Z}_0\|^2+2\|\phi(s_1+h,s)\mathbf{Z}_0-\phi(s_1,s)\mathbf{Z}_0\|^2\\
&\leqslant f(s_1,h)+C\int_{s-s_1}^0\|\mathbf{T}_\Theta(s_1+h+\theta,s)\mathbf{Z}_0-\mathbf{T}_\Theta(s_1+\theta,s)\mathbf{Z}_0\|^2d\theta,
\end{aligned}$$

where

$$\begin{aligned}
f(s_1,h):=C\Bigg\{&\|\mathbf{T}(s_1+h,s)\mathbf{Z}_0-\mathbf{T}(s_1,s)\mathbf{Z}_0\|^2\\
&+\int_{s-s_1}^0\Big|\big[\Theta(s_1+h+\theta)-\Theta(s_1+\theta)\big]\mathbf{T}_\Theta(s_1+\theta,s)\mathbf{Z}_0\Big|^2d\theta\\
&+\int_{s-s_1-h}^{s-s_1}|\Theta(s_1+h+\theta)\mathbf{T}_\Theta(s_1+h+\theta,s)\mathbf{Z}_0|^2d\theta\\
&+2\Big\|\int_{s_1}^{s_1+h}\Phi(s_1+h,\beta)\widetilde{B}(\beta)\int_s^\beta e^{D(\beta-r)}\mathcal{D}\Theta(r)\mathbf{T}_\Theta(r,s)\mathbf{Z}_0drd\beta\Big\|^2\\
&+2\Big\|\int_s^{s_1}\big[\Phi(s_1+h,\beta)-\Phi(s_1,\beta)\big]\widetilde{B}(\beta)\\
&\qquad\times\int_s^\beta e^{D(\beta-r)}\mathcal{D}\Theta(r)\mathbf{T}_\Theta(r,s)\mathbf{Z}_0drd\beta\Big\|^2\Bigg\}.
\end{aligned}$$

It follows that

$$\begin{aligned}
&\overline{\lim_{h\downarrow 0}}\|\mathbf{T}_\Theta(s_1+h,s)\mathbf{Z}_0-\mathbf{T}_\Theta(s_1,s)\mathbf{Z}_0\|^2\\
&\leqslant\overline{\lim_{h\downarrow 0}}f(s_1,h)+C\int_{s-s_1}^0\overline{\lim_{h\downarrow 0}}\|\mathbf{T}_\Theta(s_1+h+\theta,s)\mathbf{Z}_0-\mathbf{T}_\Theta(s_1+\theta,s)\mathbf{Z}_0\|^2d\theta\\
&=C\int_s^{s_1}\overline{\lim_{h\downarrow 0}}\|\mathbf{T}_\Theta(r+h,s)\mathbf{Z}_0-\mathbf{T}_\Theta(r,s)\mathbf{Z}_0\|^2dr.
\end{aligned}$$

By Gronwall's inequality, we deduce

$$\overline{\lim_{h\downarrow 0}}\|\mathbf{T}_\Theta(s_1+h,s)\mathbf{Z}_0-\mathbf{T}_\Theta(s_1,s)\mathbf{Z}_0\|^2\leqslant 0,$$

which yields

$$\lim_{h\downarrow 0}\mathbf{T}_\Theta(s_1+h,s)\mathbf{Z}_0=\mathbf{T}_\Theta(s_1,s)\mathbf{Z}_0,\quad s_1\in[s,s+\delta).\tag{2.2.24}$$

Consider $s_1\geqslant s+\delta$, there exists $r\in(s_1-\delta,s_1)$ such that

$$\lim_{t\downarrow s_1}\mathbf{T}_\Theta(t,r)\mathbf{Z}_0=\mathbf{T}_\Theta(s_1,r)\mathbf{Z}_0,$$

thus we deduce

$$\begin{aligned}&\lim_{t\downarrow s_1}\mathbf{T}_\Theta(t,s)\mathbf{Z}_0=\lim_{t\downarrow s_1}\mathbf{T}_\Theta(t,r)\mathbf{T}_\Theta(r,s)\mathbf{Z}_0\\&=\mathbf{T}_\Theta(s_1,r)\mathbf{T}_\Theta(r,s)\mathbf{Z}_0=\mathbf{T}_\Theta(s_1,s)\mathbf{Z}_0,\quad s_1\in[s+\delta,T],\end{aligned}$$

which and (2.2.24) imply that $\mathbf{T}_\Theta(t,s)\mathbf{Z}_0$ is right continuous in t on $[s,T]$. Take $h>0$, $s_2\in[s,s+\delta)$, thus

$$\begin{aligned}&\Big\|\int_s^{s_2}e^{D(s_2-r)}\mathcal{D}\Theta(r)\mathbf{T}_\Theta(r,s)\mathbf{Z}_0dr\\&\quad-\int_s^{s_2-h}e^{D(s_2-h-r)}\mathcal{D}\Theta(r)\mathbf{T}_\Theta(r,s)\mathbf{Z}_0dr\Big\|^2\\&=\int_{-\delta}^0\Big|\Theta(s_2+\theta)\mathbf{T}_\Theta(s_2+\theta,s)\mathbf{Z}_0\mathbf{1}_{[s-s_2,0]}(\theta)\\&\quad-\Theta(s_2-h+\theta)\mathbf{T}_\Theta(s_2-h+\theta,s)\mathbf{Z}_0\mathbf{1}_{[s-s_2+h,0]}(\theta)\Big|^2d\theta\\&\leqslant 3K'\int_{-\delta}^0\Big\|\big[\mathbf{T}_\Theta(s_2+\theta,s)-\mathbf{T}_\Theta(s_2-h+\theta,s)\big]\mathbf{Z}_0\mathbf{1}_{[s-s_2+h,0]}(\theta)\Big\|^2d\theta\\&\quad+3\int_{-\delta}^0\Big|\big[\Theta(s_2+\theta)-\Theta(s_2-h+\theta)\big]\mathbf{T}_\Theta(s_2-h+\theta,s)\mathbf{Z}_0\mathbf{1}_{[s-s_2+h,0]}(\theta)\Big|^2d\theta\\&\quad+3\int_{s-s_2}^{s-s_2+h}|\Theta(s_2+\theta)\mathbf{T}_\Theta(s_2+\theta,s)\mathbf{Z}_0|^2d\theta.\end{aligned}$$

Similarly, we derive

$$\lim_{h\downarrow 0}\mathbf{T}_\Theta(s_2-h,s)\mathbf{Z}_0=\mathbf{T}_\Theta(s_2,s)\mathbf{Z}_0,\quad s_2\in[s,s+\delta).\tag{2.2.25}$$

For $s_2\geqslant s+\delta$, there exists $r'\in(s_2-\delta,s_2)$ such that

$$\lim_{t\uparrow s_2}\mathbf{T}_\Theta(t,r')\mathbf{Z}_0=\mathbf{T}_\Theta(s_2,r')\mathbf{Z}_0,$$

thus we deduce

$$\begin{aligned}&\lim_{t\uparrow s_2}\mathbf{T}_\Theta(t,s)\mathbf{Z}_0=\lim_{t\uparrow s_2}\mathbf{T}_\Theta(t,r')\mathbf{T}_\Theta(r',s)\mathbf{Z}_0\\&=\mathbf{T}_\Theta(s_2,r')\mathbf{T}_\Theta(r',s)\mathbf{Z}_0=\mathbf{T}_\Theta(s_2,s)\mathbf{Z}_0,\quad s_2\in[s+\delta,T],\end{aligned}$$

which and (2.2.25) imply that $\mathbf{T}_\Theta(t,s)\mathbf{Z}_0$ is left continuous in t on $[s,T]$, thus $\mathbf{T}_\Theta(t,s)$ is strongly continuous in t on $[s,T]$. Take $h>0$, $t_1\in(t-\delta,t]$, then we have

$$\begin{aligned}
&\Big\| \int_{t_1}^{t} e^{D(t-r)}\mathcal{D}\Theta(r)\mathbf{T}_\Theta(r,t_1)\mathbf{Z}_0 dr \\
&\quad - \int_{t_1-h}^{t} e^{D(t-r)}\mathcal{D}\Theta(r)\mathbf{T}_\Theta(r,t_1-h)\mathbf{Z}_0 dr\Big\|^2 \\
&= \int_{-\delta}^{0} \Big|\Theta(t+\theta)\mathbf{T}_\Theta(t+\theta,t_1)\mathbf{Z}_0\mathbf{1}_{[t_1-t,0]}(\theta) \\
&\quad -\Theta(t+\theta)\mathbf{T}_\Theta(t+\theta,t_1-h)\mathbf{Z}_0\mathbf{1}_{[t_1-t-h,0]}(\theta)\Big|^2 d\theta \\
&\leqslant 2K'\int_{t_1-t}^{0}\Big\|\big[\mathbf{T}_\Theta(t+\theta,t_1)-\mathbf{T}_\Theta(t+\theta,t_1-h)\big]\mathbf{Z}_0\Big\|^2 d\theta \\
&\quad +2\int_{t_1-t-h}^{t_1-t} |\Theta(t+\theta)\mathbf{T}_\Theta(t+\theta,t_1-h)\mathbf{Z}_0|^2 d\theta.
\end{aligned} \tag{2.2.26}$$

As $\Theta(\cdot)$, $\mathbf{T}_\Theta(\cdot,\cdot)$ are bounded,

$$\int_{t_1-t-h}^{t_1-t} |\Theta(t+\theta)\mathbf{T}_\Theta(t+\theta,t_1-h)\mathbf{Z}_0|^2 d\theta \to 0, \quad \text{as } h\to 0. \tag{2.2.27}$$

Moreover, we have

$$\begin{aligned}
&\Big\|\int_{t_1}^{t}\int_{r}^{t} \Phi(t,\beta)\widetilde{B}(\beta)e^{D(\beta-r)}\mathcal{D}\Theta(r)\mathbf{T}_\Theta(r,t_1)\mathbf{Z}_0 d\beta dr \\
&\quad -\int_{t_1-h}^{t}\int_{r}^{t} \Phi(t,\beta)\widetilde{B}(\beta)e^{D(\beta-r)}\mathcal{D}\Theta(r)\mathbf{T}_\Theta(r,t_1-h)\mathbf{Z}_0 d\beta dr\Big\|^2 \\
&= \Big\|\int_{t_1}^{t} \Phi(t,\beta)\widetilde{B}(\beta)\int_{t_1}^{\beta} e^{D(\beta-r)}\mathcal{D}\Theta(r)\mathbf{T}_\Theta(r,t_1)\mathbf{Z}_0 dr d\beta \\
&\quad -\int_{t_1-h}^{t} \Phi(t,\beta)\widetilde{B}(\beta)\int_{t_1-h}^{\beta} e^{D(\beta-r)}\mathcal{D}\Theta(r)\mathbf{T}_\Theta(r,t_1-h)\mathbf{Z}_0 dr d\beta\Big\|^2 \\
&\leqslant 2\Big\|\int_{t_1-h}^{t_1} \Phi(t,\beta)\widetilde{B}(\beta)\int_{t_1-h}^{\beta} e^{D(\beta-r)}\mathcal{D}\Theta(r)\mathbf{T}_\Theta(r,t_1-h)\mathbf{Z}_0 dr d\beta\Big\|^2 \\
&\quad +2\Big\|\int_{t_1}^{t} \Phi(t,\beta)\widetilde{B}(\beta)\Big[\int_{t_1}^{\beta} e^{D(\beta-r)}\mathcal{D}\Theta(r)\mathbf{T}_\Theta(r,t_1)\mathbf{Z}_0 dr \\
&\quad -\int_{t_1-h}^{\beta} e^{D(\beta-r)}\mathcal{D}\Theta(r)\mathbf{T}_\Theta(r,t_1-h)\mathbf{Z}_0 dr\Big]d\beta\Big\|^2 \\
&\leqslant K'\int_{t_1-h}^{t_1}\int_{[t_1-h-\beta,0]} |\Theta(\beta+\theta)\mathbf{T}_\Theta(\beta+\theta,t_1-h)\mathbf{Z}_0|^2\mu(d\theta)d\beta
\end{aligned} \tag{2.2.28}$$

$$+C\int_{[t_1-t-h,0]}\int_{t_1\vee(t_1-h-\theta)}^{t\wedge(t_1-\theta)}|\Theta(\theta+\beta)\mathbf{T}_\Theta(\theta+\beta,t_1-h)\mathbf{Z}_0|^2d\beta\mu(d\theta)$$
$$+K\int_{[t_1-t,0]}\int_{t_1-\theta}^{t}|\Theta(\beta+\theta)(\mathbf{T}_\Theta(\beta+\theta,t_1)-\mathbf{T}_\Theta(\beta+\theta,t_1-h))\mathbf{Z}_0|^2d\beta\mu(d\theta).$$

It is easy to verify that as $h\to 0$,

$$\begin{aligned}&\int_{t_1-h}^{t_1}\int_{[t_1-h-\beta,0]}|\Theta(\beta+\theta)\mathbf{T}_\Theta(\beta+\theta,t_1-h)\mathbf{Z}_0|^2\mu(d\theta)d\beta\to 0,\\&\int_{[t_1-t-h,0]}\int_{(t_1-h-\theta)\vee t_1}^{(t_1-\theta)\wedge t}|\Theta(\theta+\beta)\mathbf{T}_\Theta(\theta+\beta,t_1-h)\mathbf{Z}_0|^2d\beta\mu(d\theta)\to 0.\end{aligned}\tag{2.2.29}$$

From (2.2.26)–(2.2.29), we deduce

$$\begin{aligned}&\|\mathbf{T}_\Theta(t,t_1)\mathbf{Z}_0-\mathbf{T}_\Theta(t,t_1-h)\mathbf{Z}_0\|^2\\&\leqslant f(t,t_1,h)+K'\int_{t_1}^{t}\|\mathbf{T}_\Theta(r,t_1)\mathbf{Z}_0-\mathbf{T}_\Theta(r,t_1-h)\mathbf{Z}_0\|^2dr,\end{aligned}$$

where

$$\begin{aligned}f(t,t_1,h):=K'\Big\{&\|\mathbf{T}(t,t_1)\mathbf{Z}_0-\mathbf{T}(t,t_1-h)\mathbf{Z}_0\|^2\\&+\int_{t_1-t-h}^{t_1-t}|\Theta(t+\theta)\mathbf{T}_\Theta(t+\theta,t_1-h)\mathbf{Z}_0|^2d\theta\\&+\int_{t_1-h}^{t_1}\int_{[t_1-h-\beta,0]}|\Theta(\beta+\theta)\mathbf{T}_\Theta(\beta+\theta,t_1-h)\mathbf{Z}_0|^2\mu(d\theta)d\beta\\&+\int_{[t_1-t-h,0]}\int_{(t_1-h-\theta)\vee t_1}^{(t_1-\theta)\wedge t}|\Theta(\theta+\beta)\mathbf{T}_\Theta(\theta+\beta,t_1-h)\mathbf{Z}_0|^2d\beta\mu(d\theta)\Big\}.\end{aligned}$$

By Gronwall's inequality, we deduce

$$\lim_{h\downarrow 0}\|\mathbf{T}_\Theta(t,t_1-h)\mathbf{Z}_0-\mathbf{T}_\Theta(t,t_1)\mathbf{Z}_0\|^2=0,$$

which yields

$$\lim_{h\downarrow 0}\mathbf{T}_\Theta(t,t_1-h)\mathbf{Z}_0=\mathbf{T}_\Theta(t,t_1)\mathbf{Z}_0,\quad t_1\in(t-\delta,t].\tag{2.2.30}$$

Consider $t_1 \leqslant t-\delta$, there exists $r'' \in (t_1, t_1+\delta)$ such that

$$\lim_{s\uparrow t_1} \mathbf{T}_\Theta(r'', s)\mathbf{Z}_0 = \mathbf{T}_\Theta(r'', t_1)\mathbf{Z}_0,$$

thus we deduce

$$\begin{aligned}\lim_{s\uparrow t_1} \mathbf{T}_\Theta(t, s)\mathbf{Z}_0 &= \lim_{s\uparrow t_1} \mathbf{T}_\Theta(t, r'')\mathbf{T}_\Theta(r'', s)\mathbf{Z}_0\\ &= \mathbf{T}_\Theta(t, r'')\mathbf{T}_\Theta(r'', t_1)\mathbf{Z}_0 = \mathbf{T}_\Theta(t, t_1)\mathbf{Z}_0, \quad t_1 \in (0, t-\delta],\end{aligned}$$

which and (2.2.30) imply that $\mathbf{T}_\Theta(t, s)\mathbf{Z}_0$ is left continuous in s on $(0, t]$. Similarly, $\mathbf{T}_\Theta(t, s)\mathbf{Z}_0$ is right continuous in s on $[0, t]$ and thus $\mathbf{T}_\Theta(t, s)$ is strongly continuous in s on $[0, t]$.

Step 5. Proof of (2.2.19). Denote

$$\begin{cases}\mathbf{T}'_0(t, s)\mathbf{Z}_0 := \mathbf{T}(t, s)\mathbf{Z}_0,\\ \mathbf{T}'_k(t, s)\mathbf{Z}_0 := \displaystyle\int_s^t \mathbf{T}'_{k-1}(t, r)\mathbf{B}\Theta(r)\mathbf{T}(r, s)\mathbf{Z}_0 dr, \quad k = 1, 2, \cdots.\end{cases} \tag{2.2.31}$$

First we aim to prove that (2.2.31) is well-defined and $\mathbf{T}'_k(t, s)\mathbf{Z}_0 = \mathbf{T}_k(t, s)\mathbf{Z}_0$. It is apparent that $\mathbf{T}'_1(t, s)\mathbf{Z}_0$ is well-defined and $\mathbf{T}'_1(t, s)\mathbf{Z}_0 = \mathbf{T}_1(t, s)\mathbf{Z}_0$. Suppose $\mathbf{T}'_k(t, s)\mathbf{Z}_0$ is well-defined and $\mathbf{T}'_k(t, s)\mathbf{Z}_0 = \mathbf{T}_k(t, s)\mathbf{Z}_0$ for all $k \leqslant l$, next we show that the conclusion holds for $k = l+1$. Noting for any $v(\cdot) \in L^2(s, T; \mathbb{R}^m)$,

$$\begin{aligned}&\int_s^t \int_r^t e^{D(t-\beta)}\mathcal{D}\Theta(\beta)\mathbf{T}(\beta, r)\mathbf{B}v(r)d\beta dr\\ &= \int_s^t e^{D(t-\beta)}\mathcal{D}\Theta(\beta)\int_s^\beta \mathbf{T}(\beta, r)\mathbf{B}v(r)dr d\beta,\end{aligned}$$

and

$$\begin{aligned}&\int_s^t \int_r^t \int_\beta^t \Phi(t, \alpha)\widetilde{B}(\alpha)e^{D(\alpha-\beta)}\mathcal{D}\Theta(\beta)\mathbf{T}(\beta, r)\mathbf{B}v(r)d\alpha d\beta dr\\ &= \int_s^t \Phi(t, \alpha)\widetilde{B}(\alpha)\int_s^\alpha e^{D(\alpha-\beta)}\mathcal{D}\Theta(\beta)\int_s^\beta \mathbf{T}(\beta, r)\mathbf{B}v(r)dr d\beta d\alpha,\end{aligned}$$

then we have

$$\begin{aligned}\int_s^t \mathbf{T}_1(t, r)\mathbf{B}v(r)dr &= \int_s^t \int_r^t \mathbf{T}(t, \beta)\mathbf{B}\Theta(\beta)\mathbf{T}(\beta, r)\mathbf{B}v(r)d\beta dr\\ &= \int_s^t \mathbf{T}(t, r)\mathbf{B}\Theta(r)\int_s^r \mathbf{T}(r, \beta)\mathbf{B}v(\beta)d\beta dr, \quad \forall v(\cdot) \in L^2(s, T; \mathbb{R}^m).\end{aligned} \tag{2.2.32}$$

It follows that

$$\begin{aligned}
\mathbf{T}'_{l+1}(t,s)\mathbf{Z}_0 &= \int_s^t \mathbf{T}'_l(t,r)\mathbf{B}\Theta(r)\mathbf{T}(r,s)\mathbf{Z}_0 dr \\
&= \int_s^t \mathbf{T}_l(t,r)\mathbf{B}\Theta(r)\mathbf{T}(r,s)\mathbf{Z}_0 dr \\
&= \int_s^t \int_r^t \mathbf{T}(t,\beta)\mathbf{B}\Theta(\beta)\mathbf{T}_{l-1}(\beta,r)\mathbf{B}\Theta(r)\mathbf{T}(r,s)\mathbf{Z}_0 d\beta dr \\
&= \int_s^t \mathbf{T}(t,\beta)\mathbf{B}\Theta(\beta)\int_s^\beta \mathbf{T}_{l-1}(\beta,r)\mathbf{B}\Theta(r)\mathbf{T}(r,s)\mathbf{Z}_0 dr d\beta \\
&= \int_s^t \mathbf{T}(t,\beta)\mathbf{B}\Theta(\beta)\mathbf{T}'_l(\beta,s)\mathbf{Z}_0 d\beta = \mathbf{T}_{l+1}(t,s)\mathbf{Z}_0,
\end{aligned}$$

which yields that (2.2.31) is well-defined and $\mathbf{T}'_l(t,s)\mathbf{Z}_0 = \mathbf{T}_l(t,s)\mathbf{Z}_0$ for $l = 1, 2, \cdots$. Therefore, by (2.2.18) we derive

$$\begin{aligned}
\mathbf{T}_\Theta(t,s)\mathbf{Z}_0 &= \sum_{k=0}^\infty \mathbf{T}_k(t,s)\mathbf{Z}_0 = \sum_{k=0}^\infty \mathbf{T}'_k(t,s)\mathbf{Z}_0 \\
&= \mathbf{T}(t,s)\mathbf{Z}_0 + \sum_{k=1}^\infty \int_s^t \mathbf{T}'_{k-1}(t,r)\mathbf{B}\Theta(r)\mathbf{T}(r,s)\mathbf{Z}_0 dr \\
&= \mathbf{T}(t,s)\mathbf{Z}_0 + \sum_{k=0}^\infty \int_s^t \mathbf{T}_k(t,r)\mathbf{B}\Theta(r)\mathbf{T}(r,s)\mathbf{Z}_0 dr \\
&= \mathbf{T}(t,s)\mathbf{Z}_0 + \int_s^t \mathbf{T}_\Theta(t,r)\mathbf{B}\Theta(r)\mathbf{T}(r,s)\mathbf{Z}_0 dr.
\end{aligned}$$

This completes the proof of Lemma 2.2.5.

□

Now we can obtain the unique solvability of (2.2.17).

Theorem 2.2.6 *Let* (H2.2)(i) *hold,* $\Theta(\cdot) : [s,T] \to \mathscr{L}(Z,\mathbb{R}^m)$ *is strongly continuous and* $\sup_{s\leqslant t\leqslant T} \|\Theta(t)\|_{\mathscr{L}(Z,\mathbb{R}^m)} < \infty$, *then there exists a unique solution to* (2.2.17)*, which is of the following form:*

$$\mathbf{Z}(t) = \mathbf{T}_\Theta(t,s)\mathbf{Z}_0 + \int_s^t \mathbf{T}_\Theta(t,r)\big[\mathbf{B}v(r) + \mathbf{b}(r)\big]dr, \quad t \in [s,T],$$

where

$$\mathbf{T}_\Theta(t,s)\mathbf{Z}_0 = \mathbf{T}(t,s)\mathbf{Z}_0 + \int_s^t \mathbf{T}(t,r)\mathbf{B}\Theta(r)\mathbf{T}_\Theta(r,s)\mathbf{Z}_0 dr, \quad t \in [s,T].$$

Proof

(i) Existence. It is sufficient to prove that

$$\begin{aligned}&\int_s^t \mathbf{T}_\Theta(t,r)\big[\mathbf{B}v(r)+\mathbf{b}(r)\big]dr\\=&\int_s^t \mathbf{T}(t,r)\Big\{\mathbf{B}\Theta(r)\int_s^r \mathbf{T}_\Theta(r,\beta)\big[\mathbf{B}v(\beta)+\mathbf{b}(\beta)\big]d\beta+\mathbf{B}v(r)+\mathbf{b}(r)\Big\}dr,\ t\in[s,T].\end{aligned}$$

Define

$$\begin{cases}\mathbf{T}_0(t,r)\mathbf{Z}_0 := \mathbf{T}(t,r)\mathbf{Z}_0,\\ \mathbf{T}_k(t,r)\mathbf{Z}_0 := \displaystyle\int_r^t \mathbf{T}(t,\beta)\mathbf{B}\Theta(\beta)\mathbf{T}_{k-1}(\beta,r)\mathbf{Z}_0 d\beta, \quad k=1,2,\cdots,\end{cases}$$

as in the proof of Lemma 2.2.5, we have

$$\mathbf{T}_k'(t,r)\mathbf{Z}_0 = \mathbf{T}_k(t,r)\mathbf{Z}_0, \quad \mathbf{T}_\Theta(t,r)\mathbf{Z}_0 = \sum_{k=0}^{\infty}\mathbf{T}_k(t,r)\mathbf{Z}_0,$$

by (2.2.18), then we only need to show that

$$\begin{aligned}&\sum_{k=0}^{\infty}\int_s^t \mathbf{T}_k(t,r)\big[\mathbf{B}v(r)+\mathbf{b}(r)\big]dr\\=&\int_s^t \mathbf{T}(t,r)\Big\{\mathbf{B}\Theta(r)\sum_{k=0}^{\infty}\int_s^r \mathbf{T}_k(r,\beta)\big[\mathbf{B}v(\beta)+\mathbf{b}(\beta)\big]d\beta\\&\quad+\mathbf{B}v(r)+\mathbf{b}(r)\Big\}dr, \quad t\in[s,T].\end{aligned} \tag{2.2.33}$$

Similar to (2.2.32), the following equality holds:

$$\begin{aligned}&\int_s^t \mathbf{T}_k(t,r)\big[\mathbf{B}v(r)+\mathbf{b}(r)\big]dr\\=&\int_s^t \mathbf{T}(t,r)\mathbf{B}\Theta(r)\int_s^r \mathbf{T}_{k-1}(r,\beta)\big[\mathbf{B}v(\beta)+\mathbf{b}(\beta)\big]d\beta dr.\end{aligned}$$

Introduce the approximation $\mathbf{B}^\lambda$ of $\mathbf{B}$ in Sect. 3.2 in advance, and noting

$$\int_s^t \mathbf{T}(t,r)\mathbf{B}^\lambda u(r)dr \to \int_s^t \mathbf{T}(t,r)\mathbf{B}u(r)dr, \text{ as } \lambda\to\infty,$$

we deduce

$$
\begin{aligned}
&\sum_{k=1}^{\infty}\int_s^t \mathbf{T}_k(t,r)\big[\mathbf{B}v(r)+\mathbf{b}(r)\big]dr\\
&=\sum_{k=1}^{\infty}\lim_{\lambda\to\infty}\int_s^t \mathbf{T}(t,r)\mathbf{B}^{\lambda}\Theta(r)\int_s^r \mathbf{T}_{k-1}(r,\beta)\big[\mathbf{B}v(\beta)+\mathbf{b}(\beta)\big]d\beta dr\\
&=\lim_{\lambda\to\infty}\sum_{k=1}^{\infty}\int_s^t \mathbf{T}(t,r)\mathbf{B}^{\lambda}\Theta(r)\int_s^r \mathbf{T}_{k-1}(r,\beta)\big[\mathbf{B}v(\beta)+\mathbf{b}(\beta)\big]d\beta dr\\
&=\lim_{\lambda\to\infty}\int_s^t \mathbf{T}(t,r)\mathbf{B}^{\lambda}\Theta(r)\sum_{k=0}^{\infty}\int_s^r \mathbf{T}_{k}(r,\beta)\big[\mathbf{B}v(\beta)+\mathbf{b}(\beta)\big]d\beta dr\\
&=\int_s^t \mathbf{T}(t,r)\mathbf{B}\Theta(r)\sum_{k=0}^{\infty}\int_s^r \mathbf{T}_{k}(r,\beta)\big[\mathbf{B}v(\beta)+\mathbf{b}(\beta)\big]d\beta dr,
\end{aligned}
$$

which implies (2.2.33).

(ii) Uniqueness. It is sufficient to show that if $\mathbf{Z}(\cdot)$ is the solution to the following integral equation:

$$
\mathbf{Z}(t)=\int_s^t \mathbf{T}(t,r)\mathbf{B}\Theta(r)\mathbf{Z}(r)dr,
$$

we necessarily have $\mathbf{Z}(t)=0$ for all $t\geqslant s$. As Step 1 in the proof of Lemma 2.2.5, there is a constant $K>0$ such that

$$
\|\mathbf{Z}(t)\|_Z^2\leqslant K\int_s^t \|\mathbf{Z}(r)\|_Z^2 dr,
$$

by Gronwall's inequality, we deduce that $\mathbf{Z}(t)=0$ for all $t\geqslant s$. This completes the proof of Theorem 2.2.6.

□

Now we can define the closed-loop solvability for Problem (P) by virtue of Problem (EP).

Definition 2.2.7 Let

$$
\begin{aligned}
\Xi[s,T]:=\Big\{&\Theta(\cdot):[s,T]\to\mathscr{L}(Z,\mathbb{R}^m)\mid\Theta(\cdot)\text{ is strongly continuous}\\
&\text{and }\sup_{s\leqslant t\leqslant T}\|\Theta(t)\|_{\mathscr{L}(Z,\mathbb{R}^m)}<\infty\Big\},
\end{aligned}
$$

and $\mathcal{Q}[s,T] := \Xi[s,T] \times L^2(s,T;\mathbb{R}^m)$. Any pair $(\Theta(\cdot), v(\cdot)) \in \mathcal{Q}[s,T]$ is called a *closed-loop strategy* of Problem (P) on $[s,T]$. For any $(\Theta(\cdot), v(\cdot)) \in \mathcal{Q}[s,T]$ and $(x, \varphi(\cdot), \psi(\cdot)) \in Z$, let $\mathbf{Z}_0 = \begin{pmatrix} x \\ \varphi \\ \psi \end{pmatrix} \in Z$, $\mathbf{Z}(\cdot) \equiv \mathbf{Z}(\cdot\,; s, \mathbf{Z}_0, \Theta(\cdot), v(\cdot))$ be the solution to (2.2.17), and let

$$u(t) = \Theta(t)\mathbf{Z}(t) + v(t), \quad t \in [s,T],$$

then $(\mathbf{Z}(\cdot), u(\cdot))$ is called the *outcome pair* of $(\Theta(\cdot), v(\cdot))$ on $[s,T]$ corresponding to the initial trajectory $(x, \varphi(\cdot), \psi(\cdot))$; $\mathbf{Z}(\cdot)$ and $u(\cdot)$ are called the corresponding *closed-loop state* and *closed-loop outcome control*, respectively.

Definition 2.2.8 A closed-loop strategy $(\Theta^*(\cdot), v^*(\cdot)) \in \mathcal{Q}[s,T]$ is said to be *optimal* on [s,T] if

$$\begin{aligned} &J(s, \mathbf{Z}_0; \Theta^*(\cdot)\mathbf{Z}^*(\cdot) + v^*(\cdot)) \leqslant J(s, \mathbf{Z}_0; \Theta(\cdot)\mathbf{Z}(\cdot) + v(\cdot)), \\ &\qquad \forall\, (\Theta(\cdot), v(\cdot)) \in \mathcal{Q}[s,T],\ \forall\, \mathbf{Z}_0 = \begin{pmatrix} x \\ \varphi \\ \psi \end{pmatrix} \in Z, \end{aligned} \tag{2.2.34}$$

where $\mathbf{Z}^*(\cdot)$ and $\mathbf{Z}(\cdot)$ are the closed-loop state corresponding to $(\Theta^*(\cdot), v^*(\cdot), x, \varphi(\cdot), \psi(\cdot))$ and $(\Theta(\cdot), v(\cdot), x, \varphi(\cdot), \psi(\cdot))$, respectively. If an optimal closed-loop strategy (uniquely) exists on $[s,T]$, Problem (P) is said to be *(uniquely) closed-loop solvable on* $[s,T]$.

Definition 2.2.9 An open-loop optimal control $u(\cdot)$ of Problem (P) is said to admit a *closed-loop representation*, if there exists a pair $(\Theta(\cdot), v(\cdot)) \in \mathcal{Q}[s,T]$ such that for any initial trajectory $(x, \varphi(\cdot), \psi(\cdot)) \in Z$, the function

$$u(t) := \Theta(t)\mathbf{Z}(t) + v(t), \quad t \in [s,T],$$

is an open-loop optimal control of Problem (P) for $(x, \varphi(\cdot), \psi(\cdot)) \in Z$, where $\mathbf{Z}(\cdot)$ is the solution to the following closed-loop system:

$$\mathbf{Z}(t) = \mathbf{T}(t,s)\mathbf{Z}_0 + \int_s^t \mathbf{T}(t,r)\big[\mathbf{B}(\Theta(r)\mathbf{Z}(r) + v(r)) + \mathbf{b}(r)\big]dr.$$

We present the following proposition.

Proposition 2.2.10 *Let* (H2.2) *hold and* $(\Theta^*(\cdot), v^*(\cdot)) \in \mathcal{Q}[s,T]$. *Then, the following three statements are equivalent:*

(i) $(\Theta^*(\cdot), v^*(\cdot))$ *is an optimal closed-loop strategy of* Problem (P) *on* $[s,T]$;
(ii) *The following inequality holds:*

$$J(s, \mathbf{Z}_0; \Theta^*(\cdot)\mathbf{Z}^*(\cdot) + v^*(\cdot)) \leqslant J(s, \mathbf{Z}_0; \Theta^*(\cdot)\mathbf{Z}(\cdot) + v(\cdot)),$$
$$\forall\, v(\cdot) \in L^2(s,T;\mathbb{R}^m),\ \forall\, \mathbf{Z}_0 = \begin{pmatrix} x \\ \varphi \\ \psi \end{pmatrix} \in Z,$$

where $\mathbf{Z}^*(\cdot)$ *and* $\mathbf{Z}(\cdot)$ *are the closed-loop state corresponding to* $(\Theta^*(\cdot), v^*(\cdot), x, \varphi(\cdot), \psi(\cdot))$ *and* $(\Theta^*(\cdot), v(\cdot), x, \varphi(\cdot), \psi(\cdot))$, *respectively.*
(iii) *The following inequality holds:*

$$J(s, \mathbf{Z}_0; \Theta^*(\cdot)\mathbf{Z}^*(\cdot) + v^*(\cdot)) \leqslant J(s, \mathbf{Z}_0; u(\cdot)),$$
$$\forall\, u(\cdot) \in L^2(s,T;\mathbb{R}^m),\ \forall\, \mathbf{Z}_0 = \begin{pmatrix} x \\ \varphi \\ \psi \end{pmatrix} \in Z,$$

where $\mathbf{Z}^*(\cdot)$ *is the closed-loop state corresponding to* $(\Theta^*(\cdot), v^*(\cdot), x, \varphi(\cdot), \psi(\cdot))$.

Proof $(i) \Rightarrow (ii)$ is trivial by taking $\Theta(\cdot) = \Theta^*(\cdot)$ in (2.2.34).

$(ii) \Rightarrow (iii)$: For any $\mathbf{Z}_0 = \begin{pmatrix} x \\ \varphi \\ \psi \end{pmatrix} \in Z$, and $u(\cdot) \in L^2(s,T;\mathbb{R}^m)$, let $\mathbf{Z}(\cdot)$ be the solution to the following integral equation:

$$\mathbf{Z}(t) = \mathbf{T}(t,s)\mathbf{Z}_0 + \int_s^t \mathbf{T}(t,r)\big[\mathbf{B}u(r) + \mathbf{b}(r)\big]dr, \tag{2.2.35}$$

and set $v(\cdot) := u(\cdot) - \Theta^*(\cdot)\mathbf{Z}(\cdot) \in L^2(s,T;\mathbb{R}^m)$, then $\mathbf{Z}(\cdot)$ is the solution to the following integral equation:

$$\mathbf{Z}(t) = \mathbf{T}(t,s)\mathbf{Z}_0 + \int_s^t \mathbf{T}(t,r)\Big\{\mathbf{B}\big[\Theta^*(r)\mathbf{Z}(r) + v(r)\big] + \mathbf{b}(r)\Big\}dr.$$

Thus it follows that

$$J(s,\mathbf{Z}_0;\Theta^*(\cdot)\mathbf{Z}^*(\cdot)+v^*(\cdot))\leqslant J(s,\mathbf{Z}_0;\Theta^*(\cdot)\mathbf{Z}(\cdot)+v(\cdot))=J(s,\mathbf{Z}_0;u(\cdot)),$$
$$\forall\, u(\cdot)\in L^2(s,T;\mathbb{R}^m),\ \forall\,\mathbf{Z}_0=\begin{pmatrix}x\\ \varphi\\ \psi\end{pmatrix}\in Z.$$

$(iii)\Rightarrow(i)$: For any $\mathbf{Z}_0=\begin{pmatrix}x\\ \varphi\\ \psi\end{pmatrix}\in Z$ and $(\Theta(\cdot),v(\cdot))\in\mathcal{Q}[s,T]$, let $\mathbf{Z}(\cdot)$ be the solution to the following equation:

$$\mathbf{Z}(t)=\mathbf{T}(t,s)\mathbf{Z}_0+\int_s^t\mathbf{T}(t,r)\Big\{\mathbf{B}\big[\Theta(r)\mathbf{Z}(r)+v(r)\big]+\mathbf{b}(r)\Big\}dr.$$

Set $u(\cdot):=\Theta(\cdot)\mathbf{Z}(\cdot)+v(\cdot)$, then $\mathbf{Z}(\cdot)$ is also the solution to (2.2.35), hence

$$J(s,\mathbf{Z}_0;\Theta^*(\cdot)\mathbf{Z}^*(\cdot)+v^*(\cdot))\leqslant J(s,\mathbf{Z}_0;u(\cdot))=J(s,\mathbf{Z}_0;\Theta(\cdot)\mathbf{Z}(\cdot)+v(\cdot)),$$
$$\forall\,(\Theta(\cdot),v(\cdot))\in\mathcal{Q}[s,T],\ \forall\,\mathbf{Z}_0=\begin{pmatrix}x\\ \varphi\\ \psi\end{pmatrix}\in Z.$$

This completes the proof of Proposition 2.2.10. □

In the end of this section, we discuss Example 1.2.4 and display lifted Problem (Ex).

Example 2.2.11 In this case, $n=m=1$, $b(t)=k_5f(t-\delta)$, $Q_{00}=q$, $R_{00}=r$, $q_0=-aq$, $Q_{10},Q_{11},S_{00},S_{11},S_{01},S_{10},R_{10},R_{11},q_1,\rho_0,\rho_1,G_{00},G_{10},G_{11},g_0,g_1=0$, and

$$\int_{-\delta}^0 A(t,\theta)\varphi(\theta)\mu(d\theta)=-k_1\varphi(0)+k_2\varphi(-\delta),$$

$$\int_{-\delta}^0 B(t,\theta)\psi(\theta)\mu(d\theta)=k_3\psi(0)+k_4\psi(-\delta).$$

The mild evolution operator $\Phi(\cdot,\cdot)$ becomes

$$\Phi(t,s):M^2\longrightarrow M^2,\quad \xi\mapsto\begin{pmatrix}\mathfrak{X}(t)\\ \mathfrak{X}_t(\cdot)\end{pmatrix},\quad \forall\,\xi:=\begin{pmatrix}x\\ \varphi\end{pmatrix}\in M^2,$$

where $\mathfrak{X}(\cdot)$ is the solution to the following equation:

$$\begin{aligned}\mathfrak{X}(t) = x + & \int_s^t \Big[-k_1\mathfrak{X}(r) + k_2\mathfrak{X}(r-\delta)\mathbf{1}_{(s+\delta,\infty)}(r)\Big]dr \\ & +k_2 \int_s^{t\wedge(s+\delta)} \varphi(r-\delta-s)dr.\end{aligned}$$

Redefine $\tilde{b}(r) := (k_5 f(r-\delta), 0)^\top$, and $\widetilde{B}(r) : L^2 \longrightarrow M^2$, $\psi(\cdot) \mapsto (k_4\psi(-\delta) + k_3\psi(0), 0)^\top$. Then the lifted state Eq. (2.1.27) becomes

$$\begin{cases} \mathbf{X}(t) = \Phi(t,s)\xi + \displaystyle\int_s^t \Phi(t,r)\big[\widetilde{B}(r)\mathbf{Y}_r + \tilde{b}(r)\big]dr, \quad t \in [s,T], \\ \mathbf{Y}_t = e^{D(t-s)}\psi + \displaystyle\int_s^t e^{D(t-r)}\mathcal{D}u(r)dr, \quad t \in [s,T]. \end{cases} \tag{2.2.36}$$

Rewrite $\mathbf{Z}_0 := \begin{pmatrix} \xi \\ \psi \end{pmatrix}$, $\mathbf{Z}(t) := \begin{pmatrix} \mathbf{X}(t) \\ \mathbf{Y}_t \end{pmatrix}$, $\mathbf{b}(t) := (k_5 f(t-\delta), 0, 0)^\top$ and $\mathbf{B} := \begin{pmatrix} 0 \\ \mathcal{D} \end{pmatrix}$. Then (2.2.36) can be written as follows

$$\mathbf{Z}(t) = \mathbf{T}(t,s)\mathbf{Z}_0 + \int_s^t \mathbf{T}(t,r)\big[\mathbf{B}u(r) + \mathbf{b}(r)\big]dr.$$

For the cost functional, redefine $\tilde{Q}_{00}(t) : \mathbb{R} \to \mathbb{R}$, $x \mapsto qx$, $\tilde{R}_{00}(t) : \mathbb{R} \to \mathbb{R}$, $y \mapsto ry$,

$$\mathbf{Q}(t) := \begin{bmatrix} \tilde{Q}_{00}(t) & 0 & 0 \\ 0 & 0 & 0 \\ 0 & 0 & 0 \end{bmatrix}, \quad \mathbf{q}(t) := \begin{bmatrix} -aq \\ 0 \\ 0 \end{bmatrix}, \quad \mathbf{R}(t) := \tilde{R}_{00}(t).$$

Then, the lifted cost (2.2.10) becomes

$$\begin{aligned} J(s,\mathbf{Z}_0; u(\cdot)) = \int_s^T & \Big[\big\langle \mathbf{Q}(t)\mathbf{Z}(t), \mathbf{Z}(t)\big\rangle_Z + \big\langle \mathbf{R}(t)u(t), u(t)\big\rangle \\ & +2\big\langle \mathbf{q}(t), \mathbf{Z}(t)\big\rangle_Z\Big]dt. \end{aligned}$$

2.3 Bibliographic Remarks

In this chapter, we apply the *product space approach* and the *extended state approach* to deal with the state delay and the control delay, respectively. We refer readers to the classical references [7–10, 13]. Among them, Delfour–Mitter [9, 10] studied the solvability of the delayed differential systems with initial data defined on $\mathbb{R}^n \times L^p$ space, and gave an integral representation of solutions, then Delfour–Mitter

[8] firstly lifted the optimal control problems with pointwise and distributed state delays to the one without delays by the *product space approach*. Later, Ichikawa [13] proposed the *extended state approach* mainly to deal with pointwise control delays. Recently, Guatteri and Masiero [12] also proposed a lifted method to address stochastic optimal control problems with general state delays, used the operator G, $B(\cdot, \cdot, \cdot)$, $\Sigma(\cdot, \cdot, \cdot)$ (see (3.2) and (3.4) in [12]) directly to expand the dimension, and lifted the finite-dimensional problems with delay to the infinite-dimensional problems without delay. Their methods could treat some nonlinear models without delays in the control.

By the product space approach and the extended state approach, the original delayed finite dimensional control problems in this book are lifted to the infinite dimensional control problems without delays. Moreover, the two control problems are equivalent. Although delays disappear in the lifted control systems, the cost is that the lifted problems have the new control operators, making the lifted state equations non-standard and the control problems cannot be handled by the existing methods in Curtain–Pritchard [4], Lasiecka–Triggiani [14, 15], Li–Yong [16]. Since $\mathbf{B}$ maps $\mathbb{R}^m$ to $M^2 \times V^*$ out Z, we can not apply the control theory in [4, 16] to address Problem (EP). [14, 15] systematically study the optimal quadratic cost problems with unbounded control operators. In their book, the unboundedness of the control operators is related to the unboundedness of the generator operators. However, after applying the *extended state approach*, the domain of the new control operators has nothing to do with the domain of the generator operators. Hence, the theory in [14, 15] is also not applicable. In the next chapter, we will use different methods to deal with the new control operators.

References

1. Bensoussan, A., Da Prato, G., Delfour, M.C., Mitter, S.K.: Representation and Control of Infinite Dimensional Systems, 2nd edn. Birkhäuser, Boston (2007)
2. Curtain, R.F.: The infinite-dimensonal Riccati equation with applications to affine hereditary differential systems. SIAM J. Control Optim. **13**, 1130–1143 (1975)
3. Curtain, R.F., Pritchard, A.J.: The infinite-dimensional Riccati equation for systems defined by the evolution operators. SIAM J. Control Optim. **14**, 951–983 (1976)
4. Curtain, R.F., Pritchard, A.J.: Infinite Dimensional Linear Systems Theory. Lecture Notes in Control and Information Sciences, vol. 8. Springer-Verlag, Berlin-New York (1978)
5. Delfour, M.C.: The linear quadratic optimal control problem for hereditary differential systems: theory and numerical solution. Appl. Math. Optim. **3**, 101–162 (1976)
6. Delfour, M.C.: State theory of linear hereditary differential systems. J. Math. Anal. Appl. **60**, 8–35 (1977)
7. Delfour, M.C., Mitter, S.K.: The optimal control of systems governed by affine hereditary differential equations. C. R. Acad. Sci. Paris, Ser. A-B **272**, A1715–A1718 (1971) (in French)
8. Delfour, M.C., Mitter, S.K.: Controllability, observability and optimal feedback control of affine hereditary differential systems. SIAM J. Control Optim. **10**, 298–328 (1972)
9. Delfour, M.C., Mitter, S.K.: Hereditary differential systems with constant delays I. General case. J. Differ. Equ. **12**, 213–235 (1972)

10. Delfour, M.C., Mitter, S.K.: Hereditary differential systems with constant delays II: A class of affine systems and the adjoint problem. J. Differ. Equ. **18**, 18–28 (1975)
11. Gibson, J.S.: Linear-quadratic optimal control of hereditary differential systems: infinite dimensional Riccati equations and numerical approximations. SIAM J. Control Optim. **21**, 95–139 (1983)
12. Guatteri, G., Masiero, F.: Stochastic maximum principle for equations with delay: going to infinite dimensions to solve the non-convex case (2023). arXiv:2306.07422
13. Ichikawa, A.: Quadratic control of evolution equations with delays in control. SIAM J. Control Optim. **20**, 645–668 (1982)
14. Lasiecka, I., Triggiani, R.: Control Theory for Partial Differential Equations: Continuous and Approximation Theories I: Abstract Parabolic Systems. Cambridge University Press, Cambridge (2000)
15. Lasiecka, I., Triggiani, R.: Control Theory for Partial Differential Equations: Continuous and Approximation Theories II: Abstract Hyperbolic-Like Systems over a Finite Time Horizon. Cambridge University Press, Cambridge (2000)
16. Li, X., Yong, J.M.: Optimal Control Theory for Infinite Dimensional Systems. Birkhäuser, Boston (1995)
17. Vinter, R.B., Kwong, R.H.: The infinite time quadratic control problem for linear systems with state and control delays: an evolution equation approach. SIAM J. Control Optim. **19**, 139–153 (1981)

Chapter 3
Solutions to the LQ Problems

In this chapter, we study the open-loop solvability, the closed-loop representation of open-loop optimal control, the closed-loop solvability and give their detailed characterizations.

3.1 Open-Loop Solvability

In this section, we first give a characterization of the open-loop solvability for Problem (P) by the lifted control Problem (EP) (see Theorems 3.1.1 and 3.1.3). However, it is not explicit due to the existence of operators, thus we specialize Radon measures to characterize the discrete delays and the distributed delays, and then give a more straightforward condition (see Theorem 3.1.7).

Theorem 3.1.1 *Let* (H2.2) *hold. For any given initial pair* $(s, x, \varphi(\cdot), \psi(\cdot)) \in [0, T) \times Z$, $u^*(\cdot)$ *is an open-loop optimal control of* Problem (P) *if and only if the following two conditions hold:*

(i) (Stationarity condition)

$$\tilde{S}_0(t)\mathbf{X}^*(t) + \tilde{R}_{01}(t)\mathbf{Y}_t^* + \tilde{R}_{00}(t)u^*(t) + [p_2(t)](0) + \tilde{\rho}_0(t) = 0, \quad \text{a.e. } t \in [s, T], \tag{3.1.1}$$

© The Author(s), under exclusive license to Springer Nature Singapore Pte Ltd. 2025
W. Meng et al., *Time-Delayed Linear Quadratic Optimal Control Problems*,
SpringerBriefs on PDEs and Data Science,
https://doi.org/10.1007/978-981-96-1897-2_3

where $(\mathbf{X}^*(\cdot), \mathbf{Y}^*_\cdot, p_1(\cdot), p_2(\cdot)) \in C([s,T];M^2)\times C([s,T];L^2)\times C([s,T];M^2)\times L^\infty(s,T;L^2)$ *is the solution to the following* forward-backward integral evolution equations *(FBIEEs, for short):*

$$\begin{cases} \mathbf{X}^*(t) = \Phi(t,s)\xi + \int_s^t \Phi(t,r)\big[\widetilde{B}(r)\mathbf{Y}^*_r + \tilde{b}(r)\big]dr, \quad t \in [s,T], \\ \mathbf{Y}^*_t = e^{D(t-s)}\psi + \int_s^t e^{D(t-r)}\mathcal{D}u^*(r)dr, \quad t \in [s,T], \\ p_1(t) = \Phi(T,t)^*\big[\tilde{G}\mathbf{X}^*(T) + \tilde{g}\big] + \int_t^T \Phi(r,t)^*\big[\tilde{Q}(r)\mathbf{X}^*(r) \\ \qquad + \tilde{S}_0(r)^*u^*(r) + \tilde{q}(r) + \tilde{S}_1(r)^*\mathbf{Y}^*_r\big]dr, \quad t \in [s,T], \\ [p_2(t)](\theta) = \int_t^{T\wedge(t+\delta+\theta)} \big[\tilde{S}_1(r)\mathbf{X}^*(r) + \tilde{R}_{01}(r)^*u^*(r) \\ \qquad + \tilde{R}_{11}(r)\mathbf{Y}^*_r + \tilde{\rho}_1(r)\big](t+\theta-r)dr \\ \qquad + \int_{[t+\theta-T,\theta]} B(t+\theta-r,r)^\top[p_1(t+\theta-r)]^0\mu(dr), \\ \qquad\qquad t \in [s,T],\ \theta \in [-\delta,0], \end{cases} \tag{3.1.2}$$

where $\xi = \begin{pmatrix} x \\ \varphi \end{pmatrix}$, $[p_1(r)]^0 \in \mathbb{R}^n$ *denotes the component of* $p_1(r)$.

(ii) (Convexity condition)

$$\begin{aligned} &\int_s^T \Big[\big\langle \tilde{Q}(t)\mathbf{X}^0(t), \mathbf{X}^0(t)\big\rangle_{M^2} + 2\big\langle \tilde{S}_0(t)\mathbf{X}^0(t), u^0(t)\big\rangle \\ &\quad + 2\big\langle \tilde{S}_1(t)\mathbf{X}^0(t), \mathbf{Y}^0_t\big\rangle_{L^2} + \big\langle \tilde{R}_{00}(t)u^0(t), u^0(t)\big\rangle \\ &\quad + 2\big\langle \tilde{R}_{01}(t)\mathbf{Y}^0_t, u^0(t)\big\rangle + \big\langle \tilde{R}_{11}(t)\mathbf{Y}^0_t, \mathbf{Y}^0_t\big\rangle_{L^2}\Big]dt \\ &\quad + \big\langle \tilde{G}\mathbf{X}^0(T), \mathbf{X}^0(T)\big\rangle_{M^2} \geqslant 0, \qquad \forall\, u^0(\cdot) \in L^2(s,T;\mathbb{R}^m), \end{aligned} \tag{3.1.3}$$

where $(\mathbf{X}^0(\cdot), \mathbf{Y}^0_\cdot)$ *is the solution to the following integral equation:*

$$\begin{cases} \mathbf{X}^0(t) = \int_s^t \Phi(t,r)\widetilde{B}(r)\mathbf{Y}^0_r dr, \quad t \in [s,T], \\ \mathbf{Y}^0_t = \int_s^t e^{D(t-r)}\mathcal{D}u^0(r)dr, \quad t \in [s,T]. \end{cases} \tag{3.1.4}$$

Proof

Step 1: we aim to prove that

$$\begin{aligned}
&\langle p_1(T), \mathbf{X}^0(T)\rangle_{M^2} + \langle p_2(T), \mathbf{Y}^0_T\rangle_{L^2} \\
&= \int_s^T \Big\{\langle p_1(t), \widetilde{B}(t)\mathbf{Y}^0_t\rangle_{M^2} - \langle \mathbf{X}^0(t), \tilde{Q}(t)\mathbf{X}^*(t) + \tilde{S}_0(t)^* u^*(t) \\
&\quad + \tilde{q}(t) + \tilde{S}_1(t)^*\mathbf{Y}^*_t\rangle_{M^2} + \langle [p_2(t)](0) - \int_{[t-T,0]} B(t-\theta,\theta)^\top \\
&\quad \times [p_1(t-\theta)]^0\mu(d\theta), u^0(t)\rangle - \langle \mathbf{Y}^0_t, \tilde{S}_1(t)\mathbf{X}^*(t) + \tilde{R}_{01}(t)^* u^*(t) \\
&\quad + \tilde{R}_{11}(t)\mathbf{Y}^*_t + \tilde{\rho}_1(t)\rangle_{L^2}\Big\} dt.
\end{aligned} \tag{3.1.5}$$

By the continuity of $B(\cdot\,,\cdot)$ and $p_1(\cdot)$, for any $\theta \in [-\delta, 0]$, $\int_{[t+\theta-T,\theta]} B(t+\theta-r, r)^\top [p_1(t+\theta-r)]^0\mu(dr)$ is meaningful, thus the last equation of (3.1.2) is well-defined. For any $(s, \xi, \psi) \in [0, T) \times Z$, suppose $(\mathbf{X}^*(\cdot), \mathbf{Y}^*_\cdot)$ and $(\mathbf{X}(\cdot), \mathbf{Y}_\cdot)$ are states corresponding to $u^*(\cdot)$ and $u(\cdot)$, respectively. For any $\varepsilon \in [0, 1]$, denote

$$\begin{aligned}
&u^\varepsilon(\cdot) := (1-\varepsilon)u^*(\cdot) + \varepsilon u(\cdot), \quad \mathbf{X}^\varepsilon(\cdot) := (1-\varepsilon)\mathbf{X}^*(\cdot) + \varepsilon \mathbf{X}(\cdot), \\
&\mathbf{Y}^\varepsilon_\cdot := (1-\varepsilon)\mathbf{Y}^*_\cdot + \varepsilon \mathbf{Y}_\cdot, \quad u^0(\cdot) := \frac{1}{\varepsilon}\big[u^\varepsilon(\cdot) - u^*(\cdot)\big] = u(\cdot) - u^*(\cdot), \\
&\mathbf{X}^0(\cdot) := \frac{1}{\varepsilon}\big[\mathbf{X}^\varepsilon(\cdot) - \mathbf{X}^*(\cdot)\big] = \mathbf{X}(\cdot) - \mathbf{X}^*(\cdot), \quad \mathbf{Y}^0_\cdot := \frac{1}{\varepsilon}\big[\mathbf{Y}^\varepsilon_\cdot - \mathbf{Y}^*_\cdot\big] = \mathbf{Y}_\cdot - \mathbf{Y}^*_\cdot.
\end{aligned}$$

Then $(\mathbf{X}^0(\cdot), \mathbf{Y}^0_\cdot, u^0(\cdot))$ satisfies the above integral Eq. (3.1.4). Therefore, it is sufficient to prove that

$$\begin{aligned}
0 = &-\int_s^T \Big\langle \widetilde{B}(t)\mathbf{Y}^0_t, \int_t^T \Phi(r,t)^*\big[\tilde{Q}(r)\mathbf{X}^*(r) + \tilde{S}_0(r)^* u^*(r) \\
&\quad + \tilde{q}(r) + \tilde{S}_1(r)^*\mathbf{Y}^*_r\big]dr\Big\rangle_{M^2} dt \\
&+ \int_s^T \langle \mathbf{X}^0(t), \tilde{Q}(t)\mathbf{X}^*(t) + \tilde{S}_0(t)^* u^*(t) + \tilde{q}(t) + \tilde{S}_1(t)^*\mathbf{Y}^*_t\rangle_{M^2} dt,
\end{aligned} \tag{3.1.6}$$

and

$$\begin{aligned}
0 = &\int_s^T \langle \mathbf{Y}^0_t, \tilde{S}_1(t)\mathbf{X}^*(t) + \tilde{R}_{01}(t)^* u^*(t) + \tilde{R}_{11}(t)\mathbf{Y}^*_t + \tilde{\rho}_1(t)\rangle_{L^2} dt \\
&- \int_s^T \Big\langle u^0(t), \int_t^{T\wedge(t+\delta)} \big[\tilde{S}_1(r)\mathbf{X}^*(r) + \tilde{R}_{01}(r)^* u^*(r) \\
&\quad + \tilde{R}_{11}(r)\mathbf{Y}^*_r + \tilde{\rho}_1(r)\big](t-r)dr\Big\rangle dt.
\end{aligned} \tag{3.1.7}$$

Using (3.1.4), (3.1.6) is easy to be verified. Then by the definition of $\mathcal{L}(t)^*$, we derive

$$
\begin{aligned}
&\int_s^T \langle \mathbf{Y}_t^0, \tilde{S}_1(t)\mathbf{X}^*(t) + \tilde{R}_{01}(t)^* u^*(t) + \tilde{R}_{11}(t)\mathbf{Y}_t^* + \tilde{\rho}_1(t)\rangle_{L^2} dt \\
&= \int_s^T \int_{-\delta}^0 \Big\langle \mathbf{Y}_t^0(\theta), \big[\tilde{S}_1(t)\mathbf{X}^*(t) + \tilde{R}_{01}(t)^* u^*(t) + \tilde{R}_{11}(t)\mathbf{Y}_t^* \\
&\qquad\qquad + \tilde{\rho}_1(t)\big](\theta)\Big\rangle d\theta dt \\
&= \int_s^{s+\delta} \int_{s-t}^0 \Big\langle u^0(t+\theta), \big[\tilde{S}_1(t)\mathbf{X}^*(t) + \tilde{R}_{01}(t)^* u^*(t) + \tilde{R}_{11}(t)\mathbf{Y}_t^* \\
&\qquad\qquad + \tilde{\rho}_1(t)\big](\theta)\Big\rangle d\theta dt \\
&\quad + \int_{s+\delta}^T \int_{-\delta}^0 \Big\langle u^0(t+\theta), \big[\tilde{S}_1(t)\mathbf{X}^*(t) + \tilde{R}_{01}(t)^* u^*(t) + \tilde{R}_{11}(t)\mathbf{Y}_t^* \\
&\qquad\qquad + \tilde{\rho}_1(t)\big](\theta)\Big\rangle d\theta dt \\
&= \int_s^{s+\delta} \int_s^t \Big\langle u^0(\theta), \big[\tilde{S}_1(t)\mathbf{X}^*(t) + \tilde{R}_{01}(t)^* u^*(t) + \tilde{R}_{11}(t)\mathbf{Y}_t^* \\
&\qquad\qquad + \tilde{\rho}_1(t)\big](\theta - t)\Big\rangle d\theta dt \\
&\quad + \int_{s+\delta}^T \int_{t-\delta}^t \Big\langle u^0(\theta), \big[\tilde{S}_1(t)\mathbf{X}^*(t) + \tilde{R}_{01}(t)^* u^*(t) + \tilde{R}_{11}(t)\mathbf{Y}_t^* \\
&\qquad\qquad + \tilde{\rho}_1(t)\big](\theta - t)\Big\rangle d\theta dt \\
&= \int_s^{s+\delta} \int_s^\theta \Big\langle u^0(t), \big[\tilde{S}_1(\theta)\mathbf{X}^*(\theta) + \tilde{R}_{01}(\theta)^* u^*(\theta) + \tilde{R}_{11}(\theta)\mathbf{Y}_\theta^* \\
&\qquad\qquad + \tilde{\rho}_1(\theta)\big](t - \theta)\Big\rangle dt d\theta \\
&\quad + \int_{s+\delta}^T \int_{\theta-\delta}^\theta \Big\langle u^0(t), \big[\tilde{S}_1(\theta)\mathbf{X}^*(\theta) + \tilde{R}_{01}(\theta)^* u^*(\theta) + \tilde{R}_{11}(\theta)\mathbf{Y}_\theta^* \\
&\qquad\qquad + \tilde{\rho}_1(\theta)\big](t - \theta)\Big\rangle dt d\theta \\
&= \int_s^T \Big\langle u^0(t), \int_t^{(t+\delta)\wedge T} \big[\tilde{S}_1(r)\mathbf{X}^*(r) + \tilde{R}_{01}(r)^* u^*(r) + \tilde{R}_{11}(r)\mathbf{Y}_r^* \\
&\qquad\qquad + \tilde{\rho}_1(r)\big](t - r) dr\Big\rangle dt,
\end{aligned}
$$

which implies (3.1.7), hence (3.1.5) holds.

Step 2: we aim to prove that

$$
\begin{aligned}
& J(s,x,\varphi(\cdot),\psi(\cdot);u^{\varepsilon}(\cdot)) - J(s,x,\varphi(\cdot),\psi(\cdot);u^{*}(\cdot)) \\
&= 2\varepsilon\Big\{\int_s^T \big\langle u^0(t), \tilde{S}_0(t)\mathbf{X}^*(t) + \tilde{R}_{01}(t)\mathbf{Y}_t^* + \tilde{R}_{00}(t)u^*(t) \\
&\quad + [p_2(t)](0) + \tilde{\rho}_0(t)\big\rangle dt\Big\} + \varepsilon^2\Big\{\int_s^T \Big[\big\langle \tilde{Q}(t)\mathbf{X}^0(t), \mathbf{X}^0(t)\big\rangle_{M^2} \\
&\quad + 2\big\langle \tilde{S}_0(t)\mathbf{X}^0(t), u^0(t)\big\rangle + 2\big\langle \tilde{S}_1(t)\mathbf{X}^0(t), \mathbf{Y}_t^0\big\rangle_{L^2} \\
&\quad + \big\langle \tilde{R}_{00}(t)u^0(t), u^0(t)\big\rangle + 2\big\langle \tilde{R}_{01}(t)\mathbf{Y}_t^0, u^0(t)\big\rangle \\
&\quad + \big\langle \tilde{R}_{11}(t)\mathbf{Y}_t^0, \mathbf{Y}_t^0\big\rangle_{L^2}\Big]dt + \big\langle \tilde{G}\mathbf{X}^0(T), \mathbf{X}^0(T)\big\rangle_{M^2}\Big\},
\end{aligned}
\tag{3.1.8}
$$

then by the arbitrariness of $u^0(\cdot)$, we complete the proof of Theorem 3.1.1. It follows that from (2.2.9),

$$
\begin{aligned}
& J(s,x,\varphi(\cdot),\psi(\cdot);u^{\varepsilon}(\cdot)) - J(s,x,\varphi(\cdot),\psi(\cdot);u^{*}(\cdot)) \\
&= 2\varepsilon\Big\{\int_s^T \Big[\big\langle u^0(t), [p_2(t)](0) + \tilde{S}_0(t)\mathbf{X}^*(t) + \tilde{R}_{00}(t)u^*(t) \\
&\qquad + \tilde{R}_{01}(t)\mathbf{Y}_t^* + \tilde{\rho}_0(t)\big\rangle + \big\langle \widetilde{B}(t)\mathbf{Y}_t^0, p_1(t)\big\rangle_{M^2}\Big]dt \\
&\qquad - \int_s^T \langle u^0(t), \int_{[t-T,0]} B(t-\theta,\theta)^\top [p_1(t-\theta)]^0 \mu(d\theta)\rangle dt\Big\} \\
&\quad + \varepsilon^2\Big\{\int_s^T \Big[\big\langle \tilde{Q}(t)\mathbf{X}^0(t), \mathbf{X}^0(t)\big\rangle_{M^2} + 2\big\langle \tilde{S}_0(t)\mathbf{X}^0(t), u^0(t)\big\rangle \\
&\qquad + 2\big\langle \tilde{S}_1(t)\mathbf{X}^0(t), \mathbf{Y}_t^0\big\rangle_{L^2} + \big\langle \tilde{R}_{00}(t)u^0(t), u^0(t)\big\rangle \\
&\qquad + 2\big\langle \tilde{R}_{01}(t)\mathbf{Y}_t^0, u^0(t)\big\rangle + \big\langle \tilde{R}_{11}(t)\mathbf{Y}_t^0, \mathbf{Y}_t^0\big\rangle_{L^2}\Big]dt \\
&\qquad + \big\langle \tilde{G}\mathbf{X}^0(T), \mathbf{X}^0(T)\big\rangle_{M^2}\Big\}.
\end{aligned}
$$

Noting

$$
\begin{aligned}
& \int_s^T \big\langle \widetilde{B}(t)\mathbf{Y}_t^0, p_1(t)\big\rangle_{M^2} dt \\
&= \int_s^T \int_{[-\delta,0]} \langle B(t,\theta)u^0(t+\theta)\mathbf{1}_{(s-t,0]}(\theta), [p_1(t)]^0\rangle \mu(d\theta)dt \\
&= \int_{[-\delta,0]} \int_{s-\theta}^T \langle B(t,\theta)u^0(t+\theta), [p_1(t)]^0\rangle dt\,\mu(d\theta) \\
&= \int_s^T \langle u^0(t), \int_{[-\delta,0]} B(t-\theta,\theta)^\top [p_1(t-\theta)]^0 \mathbf{1}_{[t-T,0]}(\theta)\mu(d\theta)\rangle dt,
\end{aligned}
$$

we deduce (3.1.8).

□

Remark 3.1.2 Noting in (3.1.2), the first two equations are the state equations of Problem (EP) and their adjoint equations are the last two equations. In other words, $p_1(\cdot)$ and $p_2(\cdot)$ are the adjoint variables of the state $\mathbf{X}^*(\cdot)$ and $\mathbf{Y}^*_\cdot$, respectively. In fact, (3.1.2) can be formalized as the following equation, in this case the equation in $p_2(\cdot)$ can be expressed by the adjoint operator e^{D^*t} of e^{Dt} and thus has the general form of adjoint equation:

$$\begin{cases} \mathbf{X}^*(t) = \Phi(t,s)\xi + \int_s^t \Phi(t,r)\big[\widetilde{B}(r)\mathbf{Y}^*_r + \tilde{b}(r)\big]dr, \ t \in [s,T], \\ \mathbf{Y}^*_t = e^{D(t-s)}\psi + \int_s^t e^{D(t-r)}\mathcal{D}u^*(r)dr, \quad t \in [s,T], \\ p_1(t) = \Phi(T,t)^*\big[\tilde{G}\mathbf{X}^*(T) + \tilde{g}\big] + \int_t^T \Phi(r,t)^*\big[\tilde{Q}(r)\mathbf{X}^*(r) \\ \qquad +\tilde{S}_0(r)^*u^*(r) + \tilde{q}(r) + \tilde{S}_1(r)^*\mathbf{Y}^*_r\big]dr, \quad t \in [s,T], \\ p_2(t) = \int_t^T e^{D^*(r-t)}\big[\tilde{S}_1(r)\mathbf{X}^*(r) + \tilde{R}_{01}(r)^*u^*(r) + \tilde{R}_{11}(r)\mathbf{Y}^*_r \\ \qquad +\tilde{\rho}_1(r)\big]dr + \int_t^T e^{D^*(r-t)}\widetilde{B}(r)^*p_1(r)dr, \quad t \in [s,T]. \end{cases} \tag{3.1.9}$$

For any $Y^1, Y^2 \in L^2$, $\int_t^T e^{D^*(r-t)}Y^1dr$ is the Bochner integral in L^2, then

$$\begin{aligned} \Big\langle \int_t^T e^{D^*(r-t)}Y^1dr, Y^2\Big\rangle_{L^2} &= \int_t^T \langle e^{D^*(r-t)}Y^1, Y^2\rangle_{L^2}dr \\ &= \int_t^T \int_{-\delta}^0 \langle [e^{D^*(r-t)}Y^1](\theta), Y^2(\theta)\rangle d\theta dr \\ &= \int_t^{T\wedge(t+\delta)} \int_{r-t-\delta}^0 \langle Y^1(t+\theta-r), Y^2(\theta)\rangle d\theta dr \\ &= \int_{-\delta}^0 \int_t^{T\wedge(t+\delta+\theta)} \langle Y^1(t+\theta-r), Y^2(\theta)\rangle dr d\theta \\ &= \int_{-\delta}^0 \Big\langle \int_t^{T\wedge(t+\delta+\theta)} Y^1(t+\theta-r)dr, Y^2(\theta)\Big\rangle d\theta \\ &= \Big\langle \int_t^{T\wedge(t+\delta+\cdot)} Y^1(t+\cdot-r)dr, Y^2\Big\rangle_{L^2}, \end{aligned}$$

thus we obtain

$$\Big[\int_t^T e^{D^*(r-t)}Y^1dr\Big](\theta) = \int_t^{T\wedge(t+\delta+\theta)} Y^1(t+\theta-r)dr, \ \theta \in [-\delta,0]. \tag{3.1.10}$$

If $\widetilde{B}(r)^* \in \mathscr{L}((M^2)^*, W^*)$, then for any $w \in W$, formally we have

$$\begin{aligned}
&\Big\langle \int_t^T e^{D^*(r-t)}\widetilde{B}(r)^* p_1(r)dr, w\Big\rangle_{W^*,W} \\
&= \int_t^T \langle e^{D^*(r-t)}\widetilde{B}(r)^* p_1(r), w\rangle_{W^*,W} dr = \int_t^T \langle p_1(r), \widetilde{B}(r)e^{D(r-t)}w\rangle_{M^2} dr \\
&= \int_t^T \Big\langle [p_1(r)]^0, \int_{[-\delta,0]} B(r,\theta)[e^{D(r-t)}w](\theta)\mu(d\theta)\Big\rangle dr \\
&= \int_{[-\delta,0]} \int_t^{T\wedge(t-\theta)} \big\langle [p_1(r)]^0, B(r,\theta)w(r-t+\theta)\big\rangle dr\,\mu(d\theta) \\
&= \int_{-\delta}^0 \int_{[-\delta,0]} \big\langle [p_1(r+t-\theta)]^0, B(r+t-\theta,\theta)w(r)\big\rangle \mathbf{1}_{[r+t-T,r]}(\theta)\mu(d\theta)dr \\
&= \int_{-\delta}^0 \Big\langle \int_{[-\delta,0]} B(t+\theta-r,r)^\top [p_1(t+\theta-r)]^0 \mathbf{1}_{[t+\theta-T,\theta]}(r)\mu(dr), w(\theta)\Big\rangle d\theta \\
&= \Big\langle \int_{[t+\cdot-T,\cdot]} B(t+\cdot-r,r)^\top [p_1(t+\cdot-r)]^0 \mu(dr), w\Big\rangle_{L^2},
\end{aligned}$$

where $[p_1(\cdot)]^0$ is the $\mathbb{R}^n$ component of $p_1(\cdot)$, thus we have

$$\begin{aligned}
&\Big[\int_t^T e^{D^*(r-t)}\widetilde{B}(r)^* p_1(r)dr\Big](\theta) \\
&= \int_{[t+\theta-T,\theta]} B(t+\theta-r,r)^\top [p_1(t+\theta-r)]^0 \mu(dr), t \in [s,T], \theta \in [-\delta,0].
\end{aligned} \tag{3.1.11}$$

By (3.1.10) and (3.1.11), we deduce (3.1.9).

Since the first two formulas of (3.1.2) can be formally written as:

$$\mathbf{Z}^*(t) = \mathbf{T}(t,s)\mathbf{Z}_0 + \int_s^t \mathbf{T}(t,r)\big[\mathbf{B}u^*(r) + \mathbf{b}(r)\big]dr, \quad t \in [s,T],$$

where $\mathbf{Z}_0 = \begin{pmatrix} x \\ \varphi \\ \psi \end{pmatrix}$, Theorem 3.1.1 can be described equivalently as follows.

Theorem 3.1.3 *Let* (H2.2) *hold. For any given initial pair* $(s, x, \varphi(\cdot), \psi(\cdot)) \in [0,T) \times Z$, $u^*(\cdot)$ *is an open-loop optimal control of* Problem (P) *if and only if the following two conditions hold:*

(i) (Stationarity condition)

$$\mathbf{S}(t)\mathbf{Z}^*(t) + \mathbf{R}(t)u^*(t) + \mathbf{B}^\top p(t) + \boldsymbol{\rho}(t) = 0, \quad \text{a.e. } t \in [s,T], \tag{3.1.12}$$

where $(\mathbf{Z}^*(\cdot), p(\cdot))$ *is the solution to the following integral equation:*

$$\begin{cases} \mathbf{Z}^*(t) = \mathbf{T}(t,s)\mathbf{Z}_0 + \int_s^t \mathbf{T}(t,r)\big[\mathbf{B}u^*(r) + \mathbf{b}(r)\big]dr, \ t \in [s,T], \\ p(t) = \mathbf{T}(T,t)^*\big[\mathbf{G}\mathbf{Z}^*(T) + \mathbf{g}\big] + \int_t^T \mathbf{T}(r,t)^*\big[\mathbf{Q}(r)\mathbf{Z}^*(r) \\ \qquad + \mathbf{S}(r)^* u^*(r) + \mathbf{q}(r)\big]dr, \quad t \in [s,T], \end{cases}$$

with $p(t) = \begin{pmatrix} [p(t)]^0 \\ [p(t)]^1 \end{pmatrix}$*. In (3.1.12),* $\mathbf{B}^\top p(t) := [p_2(t)](0)$*, and we have*

$$[p(t)]^0 = p_1(t), \qquad [p(t)]^1 = p_2(t), \quad t \in [s,T].$$

Here $(\mathbf{X}^*(\cdot), \mathbf{Y}_t^*(\cdot), p_1(\cdot), p_2(\cdot))$ *satisfies (3.1.2).*

(ii) (Convexity condition)

$$\int_s^T \big[\langle \mathbf{Q}(t)\mathbf{Z}^0(t), \mathbf{Z}^0(t)\rangle_Z + 2\langle \mathbf{S}(t)\mathbf{Z}^0(t), u^0(t)\rangle + \langle \mathbf{R}(t)u^0(t), u^0(t)\rangle\big]dt \\ + \langle \mathbf{G}\mathbf{Z}^0(T), \mathbf{Z}^0(T)\rangle_Z \geqslant 0, \qquad \forall\, u^0(\cdot) \in L^2(s,T;\mathbb{R}^m),$$

where $\mathbf{Z}^0(\cdot)$ *is the solution to the following integral equation:*

$$\mathbf{Z}^0(t) = \int_s^t \mathbf{T}(t,r)\mathbf{B}u^0(r)dr, \quad t \in [s,T].$$

Proof By (2.2.16) and (3.1.10), we have for $t \in [s,T]$,

$$[p(t)]^0 = \Phi(T,t)^*\big[\tilde{G}\mathbf{X}^*(T) + \tilde{g}\big] + \int_t^T \Phi(r,t)^*\big[\tilde{Q}(r)\mathbf{X}^*(r) \\ + \tilde{S}_0(r)^* u^*(r) + \tilde{q}(r) + \tilde{S}_1(r)^* \mathbf{Y}_r^*\big]dr = p_1(t).$$

By the continuity of $B(\cdot\,,\cdot)$ and the strong continuity of $\Phi(\cdot\,,\cdot)^*$ (see the proof of Proposition 2.2.3), each of the following terms are meaningful, and for $t \in [s,T]$, $\theta \in [-\delta, 0]$,

$$[p(t)]^1(\theta) = \Big(\int_t^T e^{D^*(r-t)}\big[\tilde{S}_1(r)\mathbf{X}^*(r) \\ + \tilde{R}_{01}(r)^* u^*(r) + \tilde{R}_{11}(r)\mathbf{Y}_r^* + \tilde{\rho}_1(r)\big]dr\Big)(\theta) \\ + \int_{[t+\theta-T,\theta]} B(t+\theta-\beta,\beta)^\top \Big\{\Phi(T,t+\theta$$

$$
\begin{aligned}
& - \beta)^*[\tilde{G}\mathbf{X}^*(T) + \tilde{g}]\Big\}^0 \mu(d\beta) \\
& + \int_t^T \int_{[t+\theta-r,\theta]} B(t+\theta-\beta,\beta)^\top \Big\{\Phi(r,t+\theta-\beta)^*\Big[\tilde{Q}(r)\mathbf{X}^*(r) \\
& + \tilde{S}_1(r)^*\mathbf{Y}_r^* + \tilde{S}_0(r)^*u^*(r) + \tilde{q}(r)\Big]\Big\}^0 \mu(d\beta)dr \\
= & \int_t^{T\wedge(t+\delta+\theta)} \big[\tilde{S}_1(r)\mathbf{X}^*(r) \\
& + \tilde{R}_{01}(r)^*u^*(r) + \tilde{R}_{11}(r)\mathbf{Y}_r^* + \tilde{\rho}_1(r)\big](t+\theta-r)dr \\
& + \int_{[t+\theta-T,\theta]} B(t+\theta-r,r)^\top[p_1(t+\theta-r)]^0\mu(dr) = p_2(t).
\end{aligned}
$$

This completes the proof of Theorem 3.1.3. □

Noting Theorem 3.1.1 characterizes the open-loop solvability for Problem (P) by lifted Problem (EP). However, it is not explicit due to the existence of operators.

In the rest of this section, we specialize $\mu(d\theta)$ to give a more straightforward characterization for the open-loop solvability of Problem (P). Consider $\mu(d\theta)$ consisting of two kinds of measures, one is defined by an integral function, for example,

$$
\mu_h(E) := \int_E h(r)dr, \qquad E \in \mathscr{B}([-\delta,0]),
$$

where non-negative function $h(\cdot) \in L^\infty(-\delta,0;\mathbb{R})$. The other is the Dirac measure, for example,

$$
\mu_\theta(E) := \begin{cases} 0, \text{ if } \theta \notin E, \\ 1, \text{ if } \theta \in E. \end{cases}
$$

Then we can express clearly the state Eq. (2.1.1) as follows:

$$
\begin{cases}
\dot{X}(t) = \sum_{i=0}^N A_i(t)X(t+\theta_i) + \int_{-\delta}^0 A^0(t,\theta)X(t+\theta)d\theta \\
\qquad + \sum_{i=0}^N B_i(t)u(t+\theta_i) + \int_{-\delta}^0 B^0(t,\theta)u(t+\theta)d\theta \\
\qquad + b(t), \quad \text{a.e. } t \in [s,T], \\
X(s) = x, \quad X(t) = \varphi(t-s), \quad t \in [s-\delta,s), \\
u(t) = \psi(t-s), \quad t \in [s-\delta,s).
\end{cases}
\tag{3.1.13}
$$

Here $A_i(\cdot)$, $A^0(\cdot,\cdot)$, $B_i(\cdot)$, $B^0(\cdot,\cdot)$ are matrix-valued functions of appropriate dimensions and $-\delta=\theta_N<\theta_{N-1}<\cdots<\theta_1<\theta_0=0$.

Similar to (2.1.10), define the following mild evolution operator (still denote $\Phi(\cdot,\cdot)$):

$$\begin{aligned}\Phi(t,s): M^2 &\longrightarrow M^2\\ \xi &\longmapsto \begin{pmatrix}\mathfrak{X}(t)\\ \mathfrak{X}_t(\cdot)\end{pmatrix}, \quad \forall\, \xi := \begin{pmatrix}x\\ \varphi\end{pmatrix}\in M^2,\end{aligned} \tag{3.1.14}$$

where $\mathfrak{X}(\cdot)$ is the solution to (3.1.13) with $B_i(\cdot), B^0(\cdot,\cdot)\equiv 0$, $i=0,\cdots,N$ and $b(\cdot)\equiv 0$. The closed linear operator $\tilde{A}(\cdot)$, associated with $\Phi(\cdot,\cdot)$, is defined as

$$\begin{aligned}\tilde{A}(t): \mathscr{W} &\longrightarrow M^2\\ \varphi &\mapsto \begin{pmatrix}\sum_{i=0}^N A_i(t)\varphi(\theta_i)+\int_{-\delta}^0 A^0(t,\theta)\varphi(\theta)d\theta\\ \dot{\varphi}(\cdot)\end{pmatrix},\end{aligned} \tag{3.1.15}$$

with

$$\mathscr{W} := \mathscr{D}(\tilde{A}(t)) = \left\{\xi = \begin{pmatrix}x\\ \varphi\end{pmatrix}\in M^2 \mid \varphi(\cdot)\in H^1(-\delta,0;\mathbb{R}^n), x=\varphi(0)\right\}.$$

$\mathscr{W}$ is dense in M^2 and is a Banach space endowed with the norm

$$\|\xi\|_{\mathscr{W}} := \|\varphi\|_{H^1}.$$

We shall denote by Λ_1 the continuous dense injection of $\mathscr{W}$ into M^2. Now (2.1.13) becomes

$$\begin{aligned}\widetilde{B}(t): L^2 &\longrightarrow M^2\\ \psi(\cdot) &\longmapsto \begin{pmatrix}\sum_{i=0}^N B_i(t)\psi(\theta_i)+\int_{-\delta}^0 B^0(t,\theta)\psi(\theta)d\theta\\ 0\end{pmatrix}.\end{aligned} \tag{3.1.16}$$

Then we can write (2.1.1) in $\mathbb{R}^n$ as the following evolution equation in M^2:

$$\begin{cases}\dot{\mathbf{X}}(t) = \tilde{A}(t)\mathbf{X}(t)+\widetilde{B}(t)u_t+\tilde{b}(t), & t\in[s,T],\\ \mathbf{X}(s)=\xi := \begin{pmatrix}x\\ \varphi\end{pmatrix}, \ u(t)=\psi(t-s), & t\in[s-\delta,s).\end{cases} \tag{3.1.17}$$

Now we use the following assumption instead of the assumption (H2.2):

(H3.1) Let $T \geqslant \delta > 0$ be given.

(i) The coefficients of the state Eq. (3.1.13) satisfy the following assumptions:

$$\begin{array}{ll} A_i(\cdot) \in C([0, T+\delta]; \mathbb{R}^{n\times n}), & A^0(\cdot\,,\cdot) \in C([0, T+\delta] \times [-\delta, 0]; \mathbb{R}^{n\times n}), \\ B_i(\cdot) \in C([0, T+\delta]; \mathbb{R}^{n\times m}), & B^0(\cdot\,,\cdot) \in C([0, T+\delta] \times [-\delta, 0]; \mathbb{R}^{n\times m}), \\ b(\cdot) \in L^2(0, T; \mathbb{R}^n), \quad i = 0, \cdots, N. & \end{array}$$

(ii) (H2.2)(ii) holds.

Furthermore we give the following results to illustrate the equivalence of (3.1.13) and (3.1.17).

Lemma 3.1.4 *Let* (H3.1)(i) *hold,* $X(\cdot\,; \cdot, \cdot, \cdot, \cdot)$ *is the solution to (3.1.13) and* $\mathbf{X}(\cdot\,; \cdot, \cdot, \cdot, \cdot)$ *is defined as follows*

$$\mathbf{X}(t; s, \xi, \psi, u) := \begin{pmatrix} X(t; s, \xi, \psi, u) \\ X_t(\cdot\,; s, \xi, \psi, u) \end{pmatrix}, \tag{3.1.18}$$

where $X_t(\theta; s, \xi, \psi, u) := X(t+\theta; s, \xi, \psi, u)$, $\theta \in [-\delta, 0]$. *Then* $\mathbf{X}(\cdot\,; \cdot, \cdot, \cdot, \cdot)$ *satisfies the following properties:*

(i) *For all* $\xi(\cdot) \in M^2$, $\psi(\cdot) \in L^2$, $u(\cdot) \in L^2(s, T; \mathbb{R}^m)$, *the map* $(t, s) \mapsto \mathbf{X}(t; s, \xi, \psi, u)$ *belongs to* $C(\Delta_*[0, T]; M^2)$. *Moreover, there exists a constant* $K > 0$ *such that for all* $(t, s) \in \Delta_*[0, T]$,

$$\|\mathbf{X}(t; s, \xi, \psi, u)\|_{M^2} \leqslant K\Big[\|\xi(\cdot)\|_{M^2} + \|u(\cdot)\|_{L^2} + \|\psi(\cdot)\|_{L^2} + \|b(\cdot)\|_{L^2}\Big].$$

(ii) *For all* $\xi(\cdot) \in \mathscr{W}$, $\psi(\cdot) \in L^2$, $u(\cdot) \in L^2(s, T; \mathbb{R}^m)$, *the map* $(t, s) \mapsto \mathbf{X}(t; s, \xi, \psi, u)$ *belongs to* $C(\Delta_*[0, T]; \mathscr{W})$. *Moreover, there exists a constant* $K > 0$ *such that for all* $(t, s) \in \Delta_*[0, T]$,

$$\|\mathbf{X}(t; s, \xi, \psi, u)\|_{\mathscr{W}} \leqslant K\Big[\|\xi(\cdot)\|_{\mathscr{W}} + \|u(\cdot)\|_{L^2} + \|\psi(\cdot)\|_{L^2} + \|b(\cdot)\|_{L^2}\Big].$$

Moreover, the map $(t, s) \mapsto D_t\mathbf{X}(t; s, \xi, \psi, u)$ *belongs to* $L^2(\Delta_*[0, T]; M^2)$, *the map* $(t, s) \mapsto D_s\mathbf{X}(t; s, \xi, \psi, u)$ *belongs to* $L^2(\Delta_*[0, T]; M^2)$, *where* D_t, D_s *indicate distributional derivatives.*

Lemma 3.1.5 *Let* (H3.1)(i) *hold.*

(i) *For all* $\xi(\cdot) \in \mathscr{W}$, $\psi(\cdot) \in L^2$, $u(\cdot) \in L^2(s, T; \mathbb{R}^m)$, $\mathbf{X}(\cdot) \equiv \mathbf{X}(\cdot\,; s, \xi, \psi, u)$ *defined as* (3.1.18), *is the unique solution to* (3.1.17) *in*

$$\mathscr{W}(s, T) := \big\{\mathbf{X}(\cdot) \in C([s, T]; \mathscr{W});\, D_t\mathbf{X}(\cdot) \in L^2(s, T; M^2)\big\}.$$

There exists another constant $K' > 0$ *such that for* (3.1.18),

$$\|\mathbf{X}(\cdot)\|_{\mathscr{W}(s,T)} \leqslant K'\Big[\|\xi(\cdot)\|_{\mathscr{W}} + \|u(\cdot)\|_{L^2} + \|\psi(\cdot)\|_{L^2} + \|b(\cdot)\|_{L^2}\Big],$$

where $\mathscr{W}(s,T)$ *is endowed with the norm:*

$$\|\mathbf{X}(\cdot)\|_{\mathscr{W}(s,T)} := \Big[\|\mathbf{X}(\cdot)\|^2_{C([s,T];\mathscr{W})} + \|D_t\mathbf{X}(\cdot)\|^2_{L^2(s,T;M^2)}\Big]^{\frac{1}{2}}. \tag{3.1.19}$$

(ii) *For all* $\xi(\cdot) \in M^2$, $\psi(\cdot) \in L^2$, $u(\cdot) \in L^2(s,T;\mathbb{R}^m)$, $\mathbf{X}(\cdot) \equiv \mathbf{X}(\cdot\,;s,\xi,\psi,u)$ *defined as* (3.1.18), *can be expressed as follows:*

$$\mathbf{X}(t) = \Phi(t,s)\xi + \int_s^t \Phi(t,r)\big[\widetilde{B}(r)u_r + \tilde{b}(r)\big]dr. \tag{3.1.20}$$

There exists a constant $K > 0$ *such that for* (3.1.18),

$$\|\mathbf{X}(\cdot)\|_{C([s,T];M^2)} \leqslant K\Big[\|\xi(\cdot)\|_{M^2} + \|u(\cdot)\|_{L^2} + \|\psi(\cdot)\|_{L^2} + \|b(\cdot)\|_{L^2}\Big].$$

Remark 3.1.6 For all $\xi \in \mathscr{W}$, suppose $X(\cdot)$ is the solution to (3.1.13), from Lemma 3.1.5 (i), $\mathbf{X}(\cdot)$ defined as (3.1.18), is the unique strong solution to (3.1.17), thus when $\xi \in \mathscr{W}$, (3.1.13) is equivalent to (3.1.17) in the sense of considering the strong solution to (3.1.17). For all $\xi \in M^2$, suppose $X(\cdot)$ is the solution to (3.1.13), from Lemma 3.1.5 (ii), $\mathbf{X}(\cdot)$ defined as (3.1.18), is the mild solution to (3.1.17), thus when $\xi \in M^2$ (3.1.13) is equivalent to (3.1.17) in the sense of considering the mild solution to (3.1.17).

Now we solve Problem (P) with (3.1.13) instead of (2.1.1) and equivalently Problem (EP) with (3.1.14)–(3.1.20) instead of (2.1.10)–(2.1.13) in Sect. 2.2. Consider the following evolution equation:

$$\begin{cases} \dot{p}_1(t) = -\Big(\tilde{A}(t)^* p_1(t) + \Lambda_1^*\big[\tilde{Q}(t)\mathbf{X}^*(t) + \tilde{S}_1(t)^*\mathbf{Y}_t^* \\ \qquad\qquad + \tilde{S}_0(t)^* u^*(t) + \tilde{q}(t)\big]\Big), \quad t \in [s,T], \\ p_1(T) = \tilde{G}\mathbf{X}^*(T) + \tilde{g}, \end{cases} \tag{3.1.21}$$

where Λ_1 is the continuous dense injection of $\mathscr{W}$ into M^2. By Theorem 4.2 in [3], in this case the third integral equation in (3.1.2) is the unique solution to (3.1.21) for given $(\mathbf{X}^*(\cdot), \mathbf{Y}^*_\cdot, u^*(\cdot))$. Suppose $p_1(t) \in M^2$ can be decomposed as $p_1(t) = \begin{pmatrix} p_1^0(t) \\ p_1^1(t) \end{pmatrix}$, where $p_1^0(t) \in \mathbb{R}^n$, $p_1^1(t) \in L^2$. Then for any $h = \begin{pmatrix} h^0 \\ h^1 \end{pmatrix} \in \mathscr{W}$, we have

$$
\begin{aligned}
&-D_t\langle p_1(t), \Lambda h\rangle_{M^2}\\
&=\big\langle \tilde{A}(t)^* p_1(t) + \Lambda_1^*[\tilde{Q}(t)\mathbf{X}^*(t) + \tilde{S}_1(t)^*\mathbf{Y}_t^* + \tilde{S}_0(t)^* u^*(t) + \tilde{q}(t)], h\big\rangle_{\mathscr{W}^*,\mathscr{W}}\\
&=\big\langle p_1(t), \tilde{A}(t)h\big\rangle_{M^2} + \big\langle \tilde{Q}(t)\mathbf{X}^*(t) + \tilde{S}_1(t)^*\mathbf{Y}_t^* + \tilde{S}_0(t)^* u^*(t) + \tilde{q}(t), h\big\rangle_{M^2}\\
&=\big\langle A_0(t)^\top p_1^0(t), h^0\big\rangle + \Big\langle \sum_{i=1}^N A_i(t)^\top p_1^0(t)\hat{\delta}(\cdot - \theta_i) + A^0(t,\cdot)^\top p_1^0(t), h^1\Big\rangle_{L^2}\\
&\quad + \Big\langle p_1^1(t), \frac{d}{d\theta}h^1\Big\rangle_{L^2} + \Big\langle Q_{00}(t)X^*(t) + \int_{-\delta}^0 Q_{10}(t,\theta)^\top X^*(t+\theta)d\theta\\
&\quad + S_{00}(t)^\top u^*(t) + q_0(t) + \int_{-\delta}^0 S_{10}(t,\theta)^\top u^*(t+\theta)d\theta, h^0\Big\rangle\\
&\quad + \Big\langle Q_{10}(t,\cdot)X^*(t) + \int_{-\delta}^0 Q_{11}(t,\theta,\cdot)X^*(t+\theta)d\theta + S_{01}(t,\cdot)^\top u^*(t)\\
&\quad + q_1(t,\cdot) + \int_{-\delta}^0 S_{11}(t,\cdot,\theta)^\top u^*(t+\theta)d\theta, h^1\Big\rangle_{L^2},
\end{aligned}
\tag{3.1.22}
$$

where $\hat{\delta}(\cdot)$ is the delta function, i.e. $\hat{\delta}(x) = 0$ for $x \neq 0$ and $\int_{-\infty}^{+\infty}\hat{\delta}(x)dx = 1$. Inspired by (3.1.22), next we introduce the following *coupled partial differential equations* (CPDEs, for short):

$$
\begin{cases}
\dot{p}_1^0(t) = -\Big(A_0(t)^\top p_1^0(t) + p_1^1(t,0) + Q_{00}(t)X^*(t)\\
\qquad + \int_{-\delta}^0 Q_{10}(t,\theta)^\top X^*(t+\theta)d\theta + S_{00}(t)^\top u^*(t)\\
\qquad + \int_{-\delta}^0 S_{10}(t,\theta)^\top u^*(t+\theta)d\theta + q_0(t)\Big), \quad \text{a.e. } t \in [s,T],\\
\Big(\frac{\partial}{\partial t} - \frac{\partial}{\partial \theta}\Big)p_1^1(t,\theta) = -\Big(\sum_{i=1}^N A_i(t)^\top p_1^0(t)\hat{\delta}(\theta - \theta_i) + A^0(t,\theta)^\top p_1^0(t)\\
\quad + Q_{10}(t,\theta)X^*(t) + \int_{-\delta}^0 Q_{11}(t,\theta',\theta)X^*(t+\theta')d\theta' + S_{01}(t,\theta)^\top u^*(t)\\
\quad + q_1(t,\theta) + \int_{-\delta}^0 S_{11}(t,\theta,\theta')^\top u^*(t+\theta')d\theta'\Big), \text{ a.e. } t \in [s,T],\ \theta \in [-\delta, 0],\\
p_1^0(T) = G_{00}X^*(T) + \int_{-\delta}^0 G_{10}(\theta)^\top X^*(T+\theta)d\theta + g_0,\\
p_1^1(t,-\delta) = 0, \quad \text{a.e. } t \in [s,T],\\
p_1^1(T,\theta) = G_{10}(\theta)X^*(T) + \int_{-\delta}^0 G_{11}(\theta',\theta)X^*(T+\theta')d\theta' + g_1(\theta),\\
\qquad\qquad\qquad\qquad \text{a.e. } \theta \in [-\delta, 0].
\end{cases}
\tag{3.1.23}
$$

Let $A_i = 0$, $i = 1, \cdots, N$, for given $(X^*(\cdot), u^*(\cdot))$, under some strict conditions, (3.1.23) admits the unique solution by the Cauchy-Kovalevskaya Theorem (see Renardy-Rogers [11]). For the general case, this type of PDEs mostly appear in the research of physics, such as the Green's function method commonly used in mathematical physics equations. If $[p_1(t)]^0 = p_1^0(t)$ and $[p_1(t)]^1 = p_1^1(t)$, then

$$
\begin{aligned}
&[p_2(t)](0) \\
&= \int_t^{T\wedge(t+\delta)} \big[\tilde{S}_1(r)\mathbf{X}^*(r) + \tilde{R}_{01}(r)^* u^*(r) + \tilde{R}_{11}(r)\mathbf{Y}_r^* + \tilde{\rho}_1(r)\big](t-r)dr \\
&\quad + \sum_{i=0}^{N} B_i(t-\theta_i)^\top p_1^0(t-\theta_i)\mathbf{1}_{[\theta_i, T-t+\theta_i)}(0) + \int_t^{T\wedge(t+\delta)} B^0(r, t-r)^\top p_1^0(r)dr \\
&= \int_0^{\delta\wedge(T-t)} \Big[S_{10}(t+\theta, -\theta)X^*(t+\theta) + \int_{-\delta}^0 S_{11}(t+\theta, \theta', -\theta)X^*(t+\theta+\theta')d\theta' \\
&\quad + R_{10}(t+\theta, -\theta)u^*(t+\theta) + \int_{-\delta}^0 R_{11}(t+\theta, \theta', -\theta)u^*(t+\theta+\theta')d\theta' \\
&\quad + \rho_1(t+\theta, -\theta) \\
&\quad + B^0(t+\theta, -\theta)^\top p_1^0(t+\theta)\Big]d\theta + \sum_{i=0}^{N} B_i(t-\theta_i)^\top p_1^0(t-\theta_i)\mathbf{1}_{[\theta_i, T-t+\theta_i]}(0).
\end{aligned}
$$

Then (3.1.1) can be written as

$$
\begin{aligned}
&S_{00}(t)X^*(t) + R_{00}(t)u^*(t) + B_0(t)^\top p_1^0(t) + \rho_0(t) \\
&+ \int_{-\delta}^0 \big[S_{01}(t,\theta)X^*(t+\theta) + R_{10}(t,\theta)^\top u^*(t+\theta)\big]d\theta \\
&+ \int_0^{\delta\wedge(T-t)} \Big[S_{10}(t+\theta, -\theta)X^*(t+\theta) \\
&+ \int_{-\delta}^0 S_{11}(t+\theta, \theta', -\theta)X^*(t+\theta+\theta')d\theta' + R_{10}(t+\theta, -\theta)u^*(t+\theta) \\
&+ \int_{-\delta}^0 R_{11}(t+\theta, \theta', -\theta)u^*(t+\theta+\theta')d\theta' + B^0(t+\theta, -\theta)^\top p_1^0(t+\theta) \\
&+ \rho_1(t+\theta, -\theta)\Big]d\theta + \sum_{i=1}^{N} B_i(t-\theta_i)^\top p_1^0(t-\theta_i)\mathbf{1}_{[\theta_i, T-t+\theta_i]}(0) = 0, \\
&\qquad\qquad\qquad\qquad\qquad\qquad \text{a.e. } t \in [s, T].
\end{aligned}
\tag{3.1.24}
$$

Now we give the following theorem to explicitly describe the open-loop solvability for Problem (P).

Theorem 3.1.7 *Let* (H3.1) *hold. For any given initial pair* $(s, x, \varphi(\cdot), \psi(\cdot)) \in [0, T) \times Z$*, assume* $u^*(\cdot)$ *is an admissible control and* $X^*(\cdot)$ *is the corresponding state. Let* $(p_1^0(t), p_1^1(t, \theta))$*,* $t \in [s, T], \theta \in [-\delta, 0]$*, satisfy (3.1.23). Suppose (3.1.24) and the following condition hold:*

$$
\begin{aligned}
&\int_s^T \Big[\big\langle Q_{00}(t)X^0(t), X^0(t)\big\rangle + 2\int_{-\delta}^0 \big\langle Q_{10}(t,\theta)^\top X^0(t+\theta), X^0(t)\big\rangle d\theta \\
&\quad + \int_{-\delta}^0\int_{-\delta}^0 \big\langle Q_{11}(t,\theta,\theta')X^0(t+\theta), X^0(t+\theta')\big\rangle d\theta' d\theta + 2\big\langle S_{00}(t)X^0(t), u^0(t)\big\rangle \\
&\quad +2\int_{-\delta}^0\int_{-\delta}^0 \big\langle S_{11}(t,\theta,\theta')X^0(t+\theta), u^0(t+\theta')\big\rangle d\theta' d\theta \\
&\quad +2\int_{-\delta}^0 \big\langle S_{01}(t,\theta)X^0(t+\theta), u^0(t)\big\rangle d\theta + 2\int_{-\delta}^0 \big\langle S_{10}(t,\theta)^\top u^0(t+\theta), X^0(t)\big\rangle d\theta \\
&\quad +\big\langle R_{00}(t)u^0(t), u^0(t)\big\rangle + 2\int_{-\delta}^0 \big\langle R_{10}(t,\theta)^\top u^0(t+\theta), u^0(t)\big\rangle d\theta \\
&\quad + \int_{-\delta}^0\int_{-\delta}^0 \big\langle R_{11}(t,\theta,\theta')u^0(t+\theta), u^0(t+\theta')\big\rangle d\theta' d\theta \Big] dt \\
&\quad +\big\langle G_{00}X^0(T), X^0(T)\big\rangle + 2\int_{-\delta}^0 \big\langle G_{10}(\theta)^\top X^0(T+\theta), X^0(T)\big\rangle d\theta \\
&\quad + \int_{-\delta}^0\int_{-\delta}^0 \big\langle G_{11}(\theta,\theta')X^0(T+\theta), X^0(T+\theta')\big\rangle d\theta' d\theta \geqslant 0, \\
&\qquad\qquad\qquad\qquad \forall\, u^0(\cdot) \in L^2(s, T; \mathbb{R}^m),
\end{aligned}
\tag{3.1.25}
$$

where $X^0(\cdot)$ *is the solution to the following ODDE:*

$$
\begin{cases}
\dot{X}^0(t) = \sum_{i=0}^N A_i(t)X^0(t+\theta_i) + \int_{-\delta}^0 A^0(t,\theta)X^0(t+\theta)d\theta \\
\qquad + \sum_{i=0}^N B_i(t)u^0(t+\theta_i) + \int_{-\delta}^0 B^0(t,\theta)u^0(t+\theta)d\theta, \ \text{a.e. } t \in [s, T], \\
X^0(t) = 0, \ u^0(t) = 0, \quad t \in [s-\delta, s].
\end{cases}
\tag{3.1.26}
$$

Then $u^*(\cdot)$ *is an open-loop optimal control of* Problem (P) *with the state Eq.* (3.1.13) *instead of* (2.1.1).

Proof For any $(s, x, \varphi(\cdot), \psi(\cdot)) \in [0, T) \times Z$, suppose $X(\cdot)$ is state corresponding to $u(\cdot)$. For any $\varepsilon \in [0, 1]$, denote

$$\begin{aligned}
u^{\varepsilon}(\cdot) &:= (1-\varepsilon)u^*(\cdot) + \varepsilon u(\cdot), \quad X^{\varepsilon}(\cdot) := (1-\varepsilon)X^*(\cdot) + \varepsilon X(\cdot),\\
u^0(\cdot) &:= \tfrac{1}{\varepsilon}\big[u^{\varepsilon}(\cdot) - u^*(\cdot)\big] = u(\cdot) - u^*(\cdot),\\
X^0(\cdot) &:= \tfrac{1}{\varepsilon}\big[X^{\varepsilon}(\cdot) - X^*(\cdot)\big] = X(\cdot) - X^*(\cdot),
\end{aligned}$$

then $(X^0(\cdot), u^0(\cdot))$ satisfies (3.1.26). Noting for a.e. $t \in [s, T]$,

$$\begin{aligned}
&\frac{d}{dt}\langle p_1^0(t), X^0(t)\rangle + \frac{d}{dt}\int_{-\delta}^{0}\langle p_1^1(t,\theta), X^0(t+\theta)\rangle d\theta\\
&= \frac{d}{dt}\langle p_1^0(t), X^0(t)\rangle + \frac{d}{dt}\int_{t-\delta}^{t}\langle p_1^1(t,r-t), X^0(r)\rangle dr\\
&= \Big\langle X^0(t), -Q_{00}(t)X^*(t) - \int_{-\delta}^{0} Q_{10}(t,\theta)^{\top}X^*(t+\theta)d\theta - S_{00}(t)^{\top}u^*(t)\\
&\quad - \int_{-\delta}^{0} S_{10}(t,\theta)^{\top}u^*(t+\theta)d\theta - q_0(t)\Big\rangle + \Big\langle p_1^0(t), B_0(t)u^0(t)\\
&\quad + \sum_{i=1}^{N} B_i(t)u^0(t+\theta_i) + \int_{-\delta}^{0} B^0(t+\theta)u^0(t+\theta)d\theta\Big\rangle\\
&\quad + \int_{t-\delta}^{t}\Big\langle X^0(r), -Q_{10}(t,r-t)X^*(t) - \int_{-\delta}^{0} Q_{11}(t,\theta',r-t)X^*(t+\theta')d\theta'\\
&\quad - S_{01}(t,r-t)^{\top}u^*(t) - q_1(t,r-t) - \int_{-\delta}^{0} S_{11}(t,r-t,\theta')^{\top}u^*(t+\theta')d\theta'\Big\rangle dr.
\end{aligned}$$

Integrating both sides of the above, we derive

$$\begin{aligned}
&J(s,x,\varphi(\cdot),\psi(\cdot);u^{\varepsilon}(\cdot)) - J(s,x,\varphi(\cdot),\psi(\cdot);u^*(\cdot))\\
&= 2\varepsilon\int_s^T\Big\{\Big\langle u^0(t), S_{00}(t)X^*(t) + R_{00}(t)u^*(t) + B_0(t)^{\top}p_1^0(t) + \rho_0(t)\\
&\qquad + \int_{-\delta}^{0} S_{01}(t,\theta)X^*(t+\theta)d\theta + \int_{-\delta}^{0} R_{10}(t,\theta)^{\top}u^*(t+\theta)d\theta\Big\rangle\\
&\qquad + \int_{-\delta}^{0}\Big\langle u^0(t+\theta), R_{10}(t,\theta)u^*(t) + S_{10}(t,\theta)X^*(t)\\
&\qquad + \int_{-\delta}^{0} S_{11}(t,\theta',\theta)X^*(t+\theta')d\theta' + \int_{-\delta}^{0} R_{11}(t,\theta',\theta)u^*(t+\theta')d\theta'\\
&\qquad + \rho_1(t,\theta) + B^0(t,\theta)^{\top}p_1^0(t) + \sum_{i=1}^{N} B_i(t)^{\top}p_1^0(t)\hat{\delta}(\theta-\theta_i)\Big\rangle d\theta\Big\}dt
\end{aligned}$$

$$
\begin{aligned}
&+\varepsilon^2\bigg\{\int_s^T\Big[2\int_{-\delta}^0\big\langle Q_{10}(t,\theta)^\top X^0(t+\theta),X^0(t)\big\rangle d\theta+\big\langle Q_{00}(t)X^0(t),X^0(t)\big\rangle\\
&\quad+\int_{-\delta}^0\int_{-\delta}^0\big\langle Q_{11}(t,\theta,\theta')X^0(t+\theta),X^0(t+\theta')\big\rangle d\theta' d\theta+2\big\langle S_{00}(t)X^0(t),\\
&\quad u^0(t)\big\rangle+2\int_{-\delta}^0\int_{-\delta}^0\big\langle S_{11}(t,\theta,\theta')X^0(t+\theta),u^0(t+\theta')\big\rangle d\theta' d\theta\\
&\quad+2\int_{-\delta}^0\big\langle S_{01}(t,\theta)X^0(t+\theta),u^0(t)\big\rangle d\theta+2\int_{-\delta}^0\big\langle R_{10}(t,\theta)^\top u^0(t+\theta),\\
&\quad u^0(t)\big\rangle d\theta+2\int_{-\delta}^0\big\langle S_{10}(t,\theta)^\top u^0(t+\theta),X^0(t)\big\rangle d\theta+\big\langle R_{00}(t)u^0(t),u^0(t)\big\rangle\\
&\quad+\int_{-\delta}^0\int_{-\delta}^0\big\langle R_{11}(t,\theta,\theta')u^0(t+\theta),u^0(t+\theta')\big\rangle d\theta' d\theta\Big]dt\\
&\quad+\big\langle G_{00}X^0(T),X^0(T)\big\rangle+2\int_{-\delta}^0\big\langle G_{10}(\theta)^\top X^0(T+\theta),X^0(T)\big\rangle d\theta\\
&\quad+\int_{-\delta}^0\int_{-\delta}^0\big\langle G_{11}(\theta,\theta')X^0(T+\theta),X^0(T+\theta')\big\rangle d\theta' d\theta\bigg\},
\end{aligned}
$$

which and (3.1.24), (3.1.25) complete the proof of Theorem 3.1.7. □

Corollary 3.1.8 *Let* (H3.1) *hold, and the state delays disappear in* Problem (P)*, i.e.* A_i, A^0, Q_{10}, Q_{11}, S_{01}, S_{11}, q_1, G_{10}, G_{11}, $g_1=0$, $i=1,\cdots,N$. *Then for any given initial pair* $(s,x,\psi(\cdot))\in[0,T)\times\mathbb{R}^n\times L^2$, $(X^*(\cdot),u^*(\cdot))$ *is the open-loop optimal pair of* Problem (P)*, with the state Eq.* (3.1.13) *instead of* (2.1.1)*, if and only if the following two conditions hold:*

(i) (Stationarity condition)

$$
\begin{aligned}
&S_{00}(t)X^*(t)+R_{00}(t)u^*(t)+B_0(t)^\top p_1^0(t)\\
&+\rho_0(t)+\int_{-\delta}^0 R_{10}(t,\theta)^\top u^*(t+\theta)d\theta\\
&+\int_0^{\delta\wedge(T-t)}\Big[S_{10}(t+\theta,-\theta)X^*(t+\theta)+R_{10}(t+\theta,-\theta)u^*(t+\theta)\\
&\qquad+\int_{-\delta}^0 R_{11}(t+\theta,\theta',-\theta)u^*(t+\theta+\theta')d\theta'\\
&\qquad+\rho_1(t+\theta,-\theta)+B^0(t+\theta,-\theta)^\top p_1^0(t+\theta)\Big]d\theta\\
&+\sum_{i=1}^N B_i(t-\theta_i)^\top p_1^0(t-\theta_i)\mathbf{1}_{[\theta_i,T-t+\theta_i]}(0)=0,\ \text{a.e. } t\in[s,T],
\end{aligned}
$$

where $(X^*(\cdot), p_1^0(\cdot))$ is the solution to the following forward-backward ordinary differential equation:

$$\begin{cases} \dot{X}^*(t) = A_0(t)X^*(t) + \sum_{i=0}^{N} B_i(t)u^*(t+\theta_i) \\ \qquad\qquad + \int_{-\delta}^{0} B^0(t,\theta)u^*(t+\theta)d\theta + b(t), \text{ a.e. } t \in [s,T], \\ \dot{p}_1^0(t) = -\Big(A_0(t)^\top p_1^0(t) + Q_{00}(t)X^*(t) + S_{00}(t)^\top u^*(t) \\ \qquad\qquad + \int_{-\delta}^{0} S_{10}(t,\theta)^\top u^*(t+\theta)d\theta + q_0(t)\Big), \text{ a.e. } t \in [s,T], \\ X^*(s) = x, \quad u^*(t) = \psi(t-s), \quad t \in [s-\delta, s], \\ p_1^0(T) = G_{00}X^*(T) + g_0. \end{cases} \tag{3.1.27}$$

(ii) (Convexity condition)

$$\begin{aligned} &\int_s^T \Big[\big\langle Q_{00}(t)X^0(t), X^0(t)\big\rangle + 2\big\langle S_{00}(t)X^0(t), u^0(t)\big\rangle \\ &\quad +2\int_{-\delta}^{0} \big\langle S_{10}(t,\theta)^\top u^0(t+\theta), X^0(t)\big\rangle d\theta + \big\langle R_{00}(t)u^0(t), u^0(t)\big\rangle \\ &\quad +2\int_{-\delta}^{0} \big\langle R_{10}(t,\theta)^\top u^0(t+\theta), u^0(t)\big\rangle d\theta \\ &\quad +\int_{-\delta}^{0}\int_{-\delta}^{0} \big\langle R_{11}(t,\theta,\theta')u^0(t+\theta), u^0(t+\theta')\big\rangle d\theta' d\theta \Big] dt \\ &\quad +\big\langle G_{00}X^0(T), X^0(T)\big\rangle \geqslant 0, \quad \forall\, u^0(\cdot) \in L^2(s,T;\mathbb{R}^m), \end{aligned}$$

where $X^0(\cdot)$ is the solution to the following ODDE:

$$\begin{cases} \dot{X}^0(t) = A_0(t)X^0(t) + \sum_{i=0}^{N} B_i(t)u^0(t+\theta_i) \\ \qquad\qquad + \int_{-\delta}^{0} B^0(t,\theta)u^0(t+\theta)d\theta, \text{ a.e. } t \in [s,T], \\ X^0(t) = 0, \; u^0(t) = 0, \quad t \in [s-\delta, s]. \end{cases}$$

Proof In this case, $p_1^1 \equiv 0$ and (3.1.23) reduces to the backward ODE in (3.1.27), and for given $(u^*(\cdot), X^*(\cdot))$, it has the unique solution. □

Remark 3.1.9 When the delays disappear in Problem (P), i.e. A_i, A^0, B_i, B^0, Q_{10}, Q_{11}, S_{01}, S_{10}, S_{11}, R_{10}, R_{11}, q_1, ρ_1, G_{10}, G_{11}, $g_1 = 0$, $i = 1, \cdots, N$, we have $p_1^1 = 0$, in this case Theorem 3.1.7 reduces to Theorem 2.3.2 in Sun-Yong [12] where the diffusion term is absent.

In the end of this section, we give the open-loop solvability of Problem (Ex).

Corollary 3.1.10 *For any given initial pair* $(s, x, \varphi(\cdot), \psi(\cdot)) \in [0, T) \times Z$, $u^*(\cdot)$ *is an open-loop optimal control of* Problem (Ex) *if and only if the following two conditions hold:*

(i) (Stationarity condition)

$$\tilde{R}_{00}(t)u^*(t) + [p_2(t)](0) = 0, \quad \text{a.e. } t \in [s, T],$$

where $(\mathbf{X}^*(\cdot), \mathbf{Y}^*_\cdot, p_1(\cdot), p_2(\cdot)) \in C([s, T]; M^2) \times C([s, T]; L^2) \times C([s, T]; M^2) \times L^\infty(s, T; L^2)$ *is the solution to the following FBIEEs:*

$$\begin{cases} \mathbf{X}^*(t) = \Phi(t, s)\xi + \int_s^t \Phi(t, r)\big[\widetilde{B}(r)\mathbf{Y}^*_r + \tilde{b}(r)\big]dr, & t \in [s, T], \\ \mathbf{Y}^*_t = e^{D(t-s)}\psi + \int_s^t e^{D(t-r)}\mathcal{D}u^*(r)dr, & t \in [s, T], \\ p_1(t) = \int_t^T \Phi(r, t)^*\big[\tilde{Q}(r)\mathbf{X}^*(r) + \tilde{q}(r)\big]dr, & t \in [s, T], \\ [p_2(t)](\theta) = k_4[p_1(t + \theta + \delta)]^0 \mathbf{1}_{[-\delta, T-t-\delta]}(\theta) \\ \qquad + k_3[p_1(t)]^0, \ t \in [s, T], \ \theta \in [-\delta, 0], \end{cases}$$

(ii) (Convexity condition)

$$\int_s^T \Big[\big\langle \tilde{Q}(t)\mathbf{X}^0(t), \mathbf{X}^0(t)\big\rangle_{M^2} + \big\langle \tilde{R}_{00}(t)u^0(t), u^0(t)\big\rangle\Big]dt \geqslant 0, \quad \forall\, u^0(\cdot) \in L^2(s, T; \mathbb{R}),$$

where $(\mathbf{X}^0(\cdot), \mathbf{Y}^0_\cdot)$ *is the solution to the following integral equation:*

$$\begin{cases} \mathbf{X}^0(t) = \int_s^t \Phi(t, r)\widetilde{B}(r)\mathbf{Y}^0_r dr, & t \in [s, T], \\ \mathbf{Y}^0_t = \int_s^t e^{D(t-r)}\mathcal{D}u^0(r)dr, & t \in [s, T]. \end{cases}$$

Corollary 3.1.11 *For any given initial pair* $(s, x, \varphi(\cdot), \psi(\cdot)) \in [0, T) \times Z$, *assume* $u^*(\cdot)$ *is an admissible control and* $X^*(\cdot)$ *is the corresponding state. Let* $(p_1^0(t), p_1^1(t,\theta))$, $t \in [s, T], \theta \in [-\delta, 0]$, *satisfy*

$$\begin{cases} \dot{p}_1^0(t) = k_1 p_1^0(t) - p_1^1(t,0) - qX^*(t), \quad \text{a.e. } t \in [s, T], \\ \left(\dfrac{\partial}{\partial t} - \dfrac{\partial}{\partial \theta}\right) p_1^1(t,\theta) = -k_2 p_1^0(t)\hat{\delta}(\theta+\delta), \text{a.e. } t \in [s, T], \ \theta \in [-\delta, 0], \\ p_1^0(T) = 0, \\ p_1^1(t, -\delta) = 0, \quad \text{a.e. } t \in [s, T], \ p_1^1(T, \theta) = 0, \text{a.e. } \theta \in [-\delta, 0]. \end{cases} \tag{3.1.28}$$

Suppose the stationarity condition

$$ru^*(t) + k_3 p_1^0(t) + k_4 p_1^0(t+\delta)\mathbf{1}_{[0,T-\delta]}(t) = 0, \ \text{a.e. } t \in [s, T], \tag{3.1.29}$$

and the convexity condition hold:

$$\int_s^T [qX^0(t)^2 + ru^0(t)^2]dt \geqslant 0, \quad \forall\, u^0(\cdot) \in L^2(s, T; \mathbb{R}), \tag{3.1.30}$$

where $X^0(\cdot)$ *is the solution to the following ODDE:*

$$\begin{cases} \dot{X}^0(t) = -k_1 X^0(t) + k_2 X^0(t-\delta) + k_3 u^0(t) + k_4 u^0(t-\delta), \ \text{a.e. } t \in [s, T], \\ X^0(t) = 0, \ u^0(t) = 0, \quad t \in [s-\delta, s]. \end{cases}$$

Then, $u^*(\cdot)$ *is an open-loop optimal control of* Problem (Ex).

By (3.1.28), we have

$$p_1^1(t, r-t) = k_2 p_1^0(r+\delta)\mathbf{1}_{[t-\delta,T-\delta]}(r),$$

then we deduce

$$\begin{cases} \dot{p}_1^0(t) = k_1 p_1^0(t) - k_2 p_1^0(t+\delta)\mathbf{1}_{[0,T-\delta]}(t) - qX^*(t), \quad \text{a.e. } t \in [s, T], \\ p_1^0(T) = 0. \end{cases}$$

It is the adjoint equation of (1.2.9) and is a type of anticipated backward ordinary differential equations, which is consistent with the results in the previous literature (see [10]).

3.2 Closed-Loop Representation of Open-Loop Optimal Control

In this section, we study the closed-loop representation of open-loop optimal control for Problem (P). Due to the existence of the new control operator $\mathbf{B}$ in (2.1.29), first we introduce the approximating problem of Problem (EP), then we try to derive the closed-loop representation of open-loop optimal control for Problem (P) by its limiting problem. However, since the range of the control operator $\mathbf{B}$ goes beyond Z, we only obtain the closed-loop representation of open-loop optimal control for Problem (P_0) (see Theorem 3.2.9). To give a more straightforward expression of the closed-loop representation of open-loop optimal control for Problem (P_0), we consider the control systems only involving the discrete delays and the distributed delays, then give a heuristic derivation and get Theorem 3.2.10.

Let $\lambda \in \rho(D)$, the resolvent set of D, and consider the approximating equation of (2.1.27):

$$\begin{cases} \mathbf{X}^\lambda(t) = \Phi(t,s)\xi + \int_s^t \Phi(t,r)\big[\widetilde{B}(r)\mathbf{Y}_r^\lambda + \tilde{b}(r)\big]dr, & t \in [s,T], \\ \mathbf{Y}_t^\lambda = e^{D(t-s)}\psi + \int_s^t e^{D(t-r)}\mathcal{D}^\lambda u(r)dr, & t \in [s,T], \end{cases} \tag{3.2.1}$$

and $\int_s^t \Phi(t,r)\tilde{B}(r)\mathbf{Y}_r^\lambda dr$ is understood in a similar way to (2.1.16), where $\mathcal{D}^\lambda := -\lambda D(\lambda I - D)^{-1}F \in \mathscr{L}(\mathbb{R}^m, L^2)$. Then (3.2.1) can be written as:

$$\mathbf{Z}^\lambda(t) = \mathbf{T}(t,s)\mathbf{Z}_0 + \int_s^t \mathbf{T}(t,r)\big[\mathbf{B}^\lambda u(r) + \mathbf{b}(r)\big]dr, \tag{3.2.2}$$

where $\mathbf{B}^\lambda := \begin{pmatrix} 0 \\ \mathcal{D}^\lambda \end{pmatrix} \in \mathscr{L}(\mathbb{R}^m, Z)$ and $\mathbf{Z}_0 := \begin{pmatrix} \xi \\ \psi \end{pmatrix}$. Noting $\lambda(\lambda I - D)^{-1} \to I$ (in the sense of strong convergence) as $\lambda \to \infty$, thus we have

$$\begin{aligned} &\int_s^t e^{D(t-r)}\mathcal{D}^\lambda u(r)dr = -\int_s^t e^{D(t-r)}\lambda D(\lambda I - D)^{-1}Fu(r)dr \\ &= -D\int_s^t e^{D(t-r)}\lambda(\lambda I - D)^{-1}Fu(r)dr \to -D\int_s^t e^{D(t-r)}Fu(r)dr, \end{aligned}$$

which implies that $\mathbf{Y}_t^\lambda \to \mathbf{Y}_t$ as $\lambda \to \infty$. Consider the following cost functional:

$$\begin{aligned} J^\lambda(s,\mathbf{Z}_0;u(\cdot)) := \int_s^T &\Big[\langle \mathbf{Q}(t)\mathbf{Z}^\lambda(t), \mathbf{Z}^\lambda(t)\rangle_Z + 2\langle \mathbf{S}(t)\mathbf{Z}^\lambda(t), u(t)\rangle \\ &+ \langle \mathbf{R}(t)u(t), u(t)\rangle + 2\langle \mathbf{q}(t), \mathbf{Z}^\lambda(t)\rangle_Z + 2\langle \boldsymbol{\rho}(t), u(t)\rangle\Big]dt \\ &+ \langle \mathbf{G}\mathbf{Z}^\lambda(T), \mathbf{Z}^\lambda(T)\rangle_Z + 2\langle \mathbf{g}, \mathbf{Z}^\lambda(T)\rangle_Z. \end{aligned}$$

Then the approximating problem of Problem (EP) can be stated as follows.

Problem ($\mathbf{EP}^\lambda$) For any $(s, \mathbf{Z}_0) \in [0, T) \times Z$, to find a $u^{\lambda *}(\cdot) \in L^2(s, T; \mathbb{R}^m)$ such that (3.2.2) is satisfied and

$$J^\lambda(s, \mathbf{Z}_0; u^{\lambda *}(\cdot)) = \inf_{u(\cdot) \in L^2(s,T;\mathbb{R}^m)} J^\lambda(s, \mathbf{Z}_0; u(\cdot)) := V^\lambda(s, \mathbf{Z}_0).$$

In the special case when $b(\cdot)$, $\mathbf{q}(\cdot)$, $\boldsymbol{\rho}(\cdot)$ and $\mathbf{g}$ vanish, denote the corresponding LQ problem, cost functional, and the value function by Problem (EP_0^λ), $J_0^\lambda(s, \mathbf{Z}_0; u(\cdot))$ and $V_0^\lambda(s, \mathbf{Z}_0)$, respectively.

The following result gives the closed-loop representation of open-loop optimal control for Problem (EP^λ).

Theorem 3.2.1 *Let* (H2.2) *hold. Suppose* $\mathbf{R} > 0$, $\mathbf{Q} - \mathbf{S}^*\mathbf{R}^{-1}\mathbf{S} \geqslant 0$, *and* $\mathbf{G} \geqslant 0$, *then for any* $z \in Z$, $P^\lambda(\cdot)$ *is the unique solution to the following integral Riccati equation in the class of strongly continuous self-adjoint operators:*

$$\begin{aligned}
(a)\quad & P^\lambda(t)z = \mathbf{T}(T,t)^*\mathbf{G}\mathbf{T}(T,t)z + \int_t^T \mathbf{T}(r,t)^*\Big\{\mathbf{Q}(r) - \big[\mathbf{B}^{\lambda *}P^\lambda(r) \\
& \quad +\mathbf{S}(r)\big]^*\mathbf{R}(r)^{-1}\big[\mathbf{B}^{\lambda *}P^\lambda(r) + \mathbf{S}(r)\big]\Big\}\mathbf{T}(r,t)z\,dr, \quad t \in [s,T], \\
(b)\quad & P^\lambda(t)z = \mathbf{T}_\Theta^\lambda(T,t)^*\mathbf{G}\mathbf{T}_\Theta^\lambda(T,t)z + \int_t^T \mathbf{T}_\Theta^\lambda(r,t)^*\Big\{\mathbf{Q}(r) \\
& \quad +\big[\mathbf{B}^{\lambda *}P^\lambda(r) + \mathbf{S}(r)\big]^*\mathbf{R}(r)^{-1}\big[\mathbf{B}^{\lambda *}P^\lambda(r) + \mathbf{S}(r)\big] \\
& \quad -\mathbf{S}(r)^*\mathbf{R}(r)^{-1}\big[\mathbf{B}^{\lambda *}P^\lambda(r) + \mathbf{S}(r)\big] \\
& \quad -\big[\mathbf{B}^{\lambda *}P^\lambda(r) + \mathbf{S}(r)\big]^*\mathbf{R}(r)^{-1}\mathbf{S}(r)\Big\}\mathbf{T}_\Theta^\lambda(r,t)z\,dr, \quad t \in [s,T], \\
(c)\quad & P^\lambda(t)z = \mathbf{T}(T,t)^*\mathbf{G}\mathbf{T}_\Theta^\lambda(T,t)z + \int_t^T \mathbf{T}(r,t)^*\Big\{\mathbf{Q}(r) \\
& \quad -\mathbf{S}(r)^*\mathbf{R}(r)^{-1}\big[\mathbf{B}^{\lambda *}P^\lambda(r) + \mathbf{S}(r)\big]\Big\}\mathbf{T}_\Theta^\lambda(r,t)z\,dr, \quad t \in [s,T],
\end{aligned} \tag{3.2.3}$$

where $\mathbf{T}_\Theta^\lambda(t,s)$ *is the perturbed mild evolution operator of* $\mathbf{T}(t,s)$ *corresponding to the perturbation* $\mathbf{B}^\lambda\Theta^\lambda(t) \equiv -\mathbf{B}^\lambda\mathbf{R}(t)^{-1}\big[\mathbf{B}^{\lambda *}P^\lambda(t) + \mathbf{S}(t)\big]$*:*

$$\mathbf{T}_\Theta^\lambda(t,s)z = \mathbf{T}(t,s)z + \int_s^t \mathbf{T}(t,r)\mathbf{B}^\lambda\Theta^\lambda(r)\mathbf{T}_\Theta^\lambda(r,s)z\,dr, \quad t \in [s,T], \tag{3.2.4}$$

and $\eta^\lambda(\cdot)$ *satisfies the following integral equation:*

$$\begin{aligned}
(a)\quad & \eta^\lambda(t) = \mathbf{T}_\Theta^\lambda(T,t)^*\mathbf{g} + \int_t^T \mathbf{T}_\Theta^\lambda(r,t)^*\Big\{\mathbf{q}(r) + P^\lambda(r)\mathbf{b}(r) \\
& \quad -\big[\mathbf{B}^{\lambda *}P^\lambda(r) + \mathbf{S}(r)\big]^*\mathbf{R}(r)^{-1}\boldsymbol{\rho}(r)\Big\}dr, \quad t \in [s,T], \\
(b)\quad & \eta^\lambda(t) = \mathbf{T}(T,t)^*\mathbf{g} + \int_t^T \mathbf{T}(r,t)^*\Big\{P^\lambda(r)\mathbf{b}(r) - \big[\mathbf{B}^{\lambda *}P^\lambda(r) \\
& \quad +\mathbf{S}(r)\big]^*\mathbf{R}(r)^{-1}\boldsymbol{\rho}(r) - \big[\mathbf{B}^{\lambda *}P^\lambda(r) + \mathbf{S}(r)\big]^* \\
& \quad \times\mathbf{R}(r)^{-1}\mathbf{B}^{\lambda *}\eta^\lambda(r) + \mathbf{q}(r)\Big\}dr, \quad t \in [s,T].
\end{aligned} \tag{3.2.5}$$

Moreover, the optimal control of Problem (P^λ) *is*

$$u^{\lambda*}(t) = \Theta^\lambda(t)\mathbf{Z}^{\lambda*}(t) - \mathbf{R}(t)^{-1}\big[\mathbf{B}^{\lambda^*}\eta^\lambda(t) + \rho(t)\big], \quad \text{a.e. } t \in [s, T], \tag{3.2.6}$$

and the value function is

$$\begin{aligned} V^\lambda(s, \mathbf{Z}_0) &= \big\langle P^\lambda(s)\mathbf{Z}_0, \mathbf{Z}_0\big\rangle_Z + 2\big\langle\eta^\lambda(s), \mathbf{Z}_0\big\rangle_Z \\ &+ \int_s^T \Big\{2\big\langle\eta^\lambda(t), \mathbf{b}(t)\big\rangle_Z - \big\langle\mathbf{R}(t)^{-1}\big[\mathbf{B}^{\lambda^*}\eta^\lambda(t) \\ &\qquad +\rho(t)\big], \mathbf{B}^{\lambda^*}\eta^\lambda(t) + \rho(t)\big\rangle\Big\}dt. \end{aligned} \tag{3.2.7}$$

Proof We introduce the following auxiliary optimal control problem:

$$\begin{cases} \min\limits_{\tilde{u}(\cdot)} \tilde{J}^\lambda(s, \mathbf{Z}_0; \tilde{u}(\cdot)) := \displaystyle\int_s^T \Big\{\big\langle[\mathbf{Q}(t) - \mathbf{S}(t)^*\mathbf{R}(t)^{-1}\mathbf{S}(t)]\tilde{\mathbf{Z}}^\lambda(t), \\ \qquad\qquad \tilde{\mathbf{Z}}^\lambda(t)\big\rangle_Z + \big\langle\mathbf{R}(t)\tilde{u}(t), \tilde{u}(t)\big\rangle\Big\}dt + \big\langle\mathbf{G}\tilde{\mathbf{Z}}^\lambda(T), \tilde{\mathbf{Z}}^\lambda(T)\big\rangle_Z, \\ \text{subject to} \qquad \tilde{\mathbf{Z}}^\lambda(t) = \tilde{\mathbf{T}}^\lambda(t, s)\mathbf{Z}_0 + \displaystyle\int_s^t \tilde{\mathbf{T}}^\lambda(t, r)\mathbf{B}^\lambda\tilde{u}(r)dr, \\ \tilde{u}(\cdot) \in L^2(s, T; \mathbb{R}^m), \end{cases}$$

where $\tilde{\mathbf{T}}^\lambda(t, s)$ is the perturbed mild evolution operator of $\mathbf{T}(t, s)$ corresponding to the perturbation $-\mathbf{B}^\lambda\mathbf{R}(t)^{-1}\mathbf{S}(t)$, denote the above problem by Problem ($\tilde{P}_0^\lambda$). It is easy to verify that

$$\begin{pmatrix} \mathbf{Z}^\lambda \\ u \end{pmatrix} = \begin{pmatrix} \tilde{\mathbf{Z}}^\lambda \\ -\mathbf{R}^{-1}\mathbf{S}\tilde{\mathbf{Z}}^\lambda + \tilde{u} \end{pmatrix} = \begin{pmatrix} I & 0 \\ -\mathbf{R}^{-1}\mathbf{S} & I \end{pmatrix} \begin{pmatrix} \tilde{\mathbf{Z}}^\lambda \\ \tilde{u} \end{pmatrix}, \tag{3.2.8}$$

thus Problem (P_0^λ) is equivalent to Problem ($\tilde{P}_0^\lambda$). By Theorem 9.18 in [2], there exists the optimal control

$$\bar{\tilde{u}}^\lambda(t) = -\mathbf{R}(t)^{-1}\mathbf{B}^{\lambda^*}\tilde{P}^\lambda(t)\bar{\tilde{\mathbf{Z}}}^\lambda(t), \quad \text{a.e. } t \in [s, T], \tag{3.2.9}$$

where $\tilde{P}^\lambda(t)$ is the unique solution to the integral Riccati equation:

$$\begin{aligned} \tilde{P}^\lambda(t)z = \tilde{\mathbf{T}}_\Theta^\lambda(T, t)^*\mathbf{G}\tilde{\mathbf{T}}_\Theta^\lambda(T, t)z + \int_t^T \tilde{\mathbf{T}}_\Theta^\lambda(r, t)^*\Big\{\mathbf{Q}(r) \\ -\mathbf{S}(r)^*\mathbf{R}(r)^{-1}\mathbf{S}(r) + \tilde{P}^\lambda(r)\mathbf{B}^\lambda\mathbf{R}(r)^{-1}\mathbf{B}^{\lambda^*}\tilde{P}^\lambda(r)\Big\} \\ \times\tilde{\mathbf{T}}_\Theta^\lambda(r, t)zdr, \quad t \in [s, T], \quad z \in Z, \end{aligned} \tag{3.2.10}$$

where $\tilde{\mathbf{T}}^{\lambda}_{\Theta}(t,s)$ is the perturbed mild evolution operator of $\tilde{\mathbf{T}}^{\lambda}(t,s)$ corresponding to the perturbation $-\mathbf{B}^{\lambda}\mathbf{R}(t)^{-1}\mathbf{B}^{\lambda *}\tilde{P}^{\lambda}(t)$, i.e., $\tilde{\mathbf{T}}^{\lambda}_{\Theta}(t,s)$ satisfies the following integral equation:

$$\tilde{\mathbf{T}}^{\lambda}_{\Theta}(t,s)z = \tilde{\mathbf{T}}^{\lambda}(t,s)z - \int_s^t \tilde{\mathbf{T}}^{\lambda}(t,r)\mathbf{B}^{\lambda}\mathbf{R}(r)^{-1}\mathbf{B}^{\lambda *}\tilde{P}^{\lambda}(r)\tilde{\mathbf{T}}^{\lambda}_{\Theta}(r,s)zdr,$$

and the optimal cost is given by

$$\tilde{V}(s,\mathbf{Z}_0) = \langle \tilde{P}^{\lambda}(s)\mathbf{Z}_0, \mathbf{Z}_0 \rangle.$$

Noting $\tilde{\mathbf{T}}^{\lambda}_{\Theta}(t,s)$ also satisfies the following integral equation:

$$\tilde{\mathbf{T}}^{\lambda}_{\Theta}(t,s)z = \mathbf{T}(t,s)z - \int_s^t \mathbf{T}(t,r)\mathbf{B}^{\lambda}\mathbf{R}(r)^{-1}\big[\mathbf{B}^{\lambda *}\tilde{P}^{\lambda}(r) + \mathbf{S}(r)\big]\tilde{\mathbf{T}}^{\lambda}_{\Theta}(r,s)zdr,$$

and (3.2.10) can be written as

$$\begin{aligned}\tilde{P}^{\lambda}(t)z = {} & \tilde{\mathbf{T}}^{\lambda}_{\Theta}(T,t)^*\mathbf{G}\tilde{\mathbf{T}}^{\lambda}_{\Theta}(T,t)z + \int_t^T \tilde{\mathbf{T}}^{\lambda}_{\Theta}(r,t)^*\Big\{\mathbf{Q}(r) \\ & -\mathbf{S}(r)^*\mathbf{R}(r)^{-1}\big[\mathbf{S}(r) + \mathbf{B}^{\lambda *}\tilde{P}^{\lambda}(r)\big] - \big[\tilde{P}^{\lambda}(r)\mathbf{B}^{\lambda} \\ & +\mathbf{S}(r)^*\big]\mathbf{R}(r)^{-1}\mathbf{S}(r) + \big[\tilde{P}^{\lambda}(r)\mathbf{B}^{\lambda} + \mathbf{S}(r)^*\big]\mathbf{R}(r)^{-1} \\ & \times\big[\mathbf{B}^{\lambda *}\tilde{P}^{\lambda}(r) + \mathbf{S}(r)\big]\Big\}\tilde{\mathbf{T}}^{\lambda}_{\Theta}(r,t)zdr, \quad t \in [s,T], \quad z \in Z,\end{aligned}$$

thus $P^{\lambda}(t) = \tilde{P}^{\lambda}(t)$ and $\mathbf{T}^{\lambda}_{\Theta}(t,s) = \tilde{\mathbf{T}}^{\lambda}_{\Theta}(t,s)$. From (3.2.8) and (3.2.9), we deduce that

$$\bar{u}^{\lambda}(t) = -\mathbf{R}(t)^{-1}\big[\mathbf{B}^{\lambda *}P^{\lambda}(t) + \mathbf{S}(t)\big]\bar{\mathbf{Z}}^{\lambda}(t),$$

is the optimal control of Problem (P_0^{λ}). Let $\eta^{\lambda}(\cdot)$ be the solution to (3.2.5) and using the method of completion of squares we can show that (3.2.6) is the optimal control of Problem (P^{λ}) and (3.2.7) is the value function. Because the completion of squares technique is standard and similar proof will appear in Theorem 3.3.7 in Sect. 3.3, so we omit the detailed proof here. The equivalence of the three integral equations in (3.2.3) and the equivalence of the two equations in (3.2.5) can be proved using the similar method in Lemma 3.3.8 in Sect. 3.3. Noting from Theorem 9.19 in [2], $P^{\lambda}(t)$ is weakly continuous. However, by Proposition 1.3.6 $\mathbf{T}^{\lambda}_{\Theta}(t,s)$ is a strongly continuous mild evolution operator, recalling Proposition 2.2.3, by the strong continuity of $\mathbf{T}(t, s)^*$ and (3.2.3)(c) we deduce that $P^\lambda(t)$ is strongly continuous. This completes the proof of Theorem 3.2.1. □

Corollary 3.2.2 *Let* (H2.2) *hold. Suppose* $\mathbf{R} > 0$, $\mathbf{Q} - \mathbf{S}^*\mathbf{R}^{-1}\mathbf{S} \geqslant 0$, *and* $\mathbf{G} \geqslant 0$, *then for any* $z \in Z$, $P^\lambda(\cdot)$ *is the solution to the integral Riccati equation (3.2.3). Furthermore the optimal control of* Problem (P_0^λ) *is*

$$\bar{u}^\lambda(t) = \Theta^\lambda(t)\bar{\mathbf{Z}}^\lambda(t), \quad \text{a.e. } t \in [s, T], \tag{3.2.11}$$

where $\Theta^\lambda(t) := -\mathbf{R}(t)^{-1}\big[\mathbf{B}^{\lambda *} P^\lambda(t) + \mathbf{S}(t)\big]$, *and the value function is*

$$V_0^\lambda(s, \mathbf{Z}_0) = \big\langle P^\lambda(s)\mathbf{Z}_0, \mathbf{Z}_0\big\rangle_Z.$$

Remark 3.2.3 Regardless of the lifting process, if we consider alone Problem (P^λ), which is a general infinite dimensional LQ optimal control problem with bounded control operator, Theorem 3.2.1 gives its closed-loop representation of open-loop optimal control. Although there have been a lot of related references, in which either the problems are standard, that is the state equations are only homogeneous and the cost functionals are without the cross terms of state and control, linear terms of state and control, or the problems are other special cases, Theorem 3.2.1 gives a more general result, which plays an important role in the study of the differential games. Moreover, under some mild conditions, P^λ is strongly continuous, which is crucial in the study of the closed-loop solvability in Sect. 3.3.

Next we study the limit Problem (EP_0^λ) to derive the closed-loop representation of open-loop optimal control for Problem (P_0). For convenience, we restate Problem (P_0) as follows.

Problem $(\boldsymbol{P_0})$ For any $(s, x, \varphi(\cdot), \psi(\cdot)) \in [0, T) \times Z$, to find a $u^*(\cdot) \in L^2(s, T; \mathbb{R}^m)$ such that (2.1.1) is satisfied with $b(\cdot) \equiv 0$ and

$$\begin{aligned} &J_0(s, x, \varphi(\cdot), \psi(\cdot); u^*(\cdot)) \\ &= \inf_{u(\cdot) \in L^2(s,T;\mathbb{R}^m)} J_0(s, x, \varphi(\cdot), \psi(\cdot); u(\cdot)) := V_0(s, x, \varphi(\cdot), \psi(\cdot)), \end{aligned}$$

where $J_0(s, x, \varphi(\cdot), \psi(\cdot); u(\cdot))$ is defined by (2.2.1), with $q_0(\cdot)$, $q_1(\cdot\,,\cdot)$, $\rho_0(\cdot)$, $\rho_1(\cdot\,,\cdot)$, g_0 and $g_1(\cdot)$ vanishing.

Now introduce the following hypotheses.

(H3.2) Let $T \geqslant \delta > 0$ be given.

(i) (H2.2) holds.
(ii) The weight coefficients of the cost functional (2.2.1) are all continuous.
(iii) $R_{00} > 0$, $S_{00}, S_{01}, R_{10} = 0$, $\mathbf{Q} \geqslant 0$, $\mathbf{G} \geqslant 0$.

Lemma 3.2.4 *Suppose that $u^\lambda(\cdot)$ converges weakly to $u(\cdot)$ in $L^2(s, T; \mathbb{R}^m)$, then*

$$\mathbf{Y}_t^\lambda = e^{D(t-s)}\psi + \int_s^t e^{D(t-r)}\mathcal{D}^\lambda u^\lambda(r)dr,$$

converges weakly to $\mathbf{Y}_t$ given by

$$\mathbf{Y}_t = e^{D(t-s)}\psi + \int_s^t e^{D(t-r)}\mathcal{D}u(r)dr, \quad \text{for all } t \in [s, T].$$

Lemma 3.2.5 *If $\mathbf{Y} \in L^2$ and $\mathbf{Y}(\cdot)$ is continuous at 0, then $(\mathcal{D}^\lambda)^*\mathbf{Y} \to \mathbf{Y}(0)$ as $\lambda \to \infty$.*

Recalling (2.1.25), notice that $\mathcal{D}^* \in \mathscr{L}(V, \mathbb{R}^m)$ and $\mathcal{D}^* w = w(0)$ for any $w \in V$, then from the above lemma $\mathcal{D}^*\mathbf{Y}$ can be defined for $\mathbf{Y} \in L^2$ which is continuous at 0, thus $\mathcal{D}^*$ can be defined in the generalized sense.

Lemma 3.2.6 *Let* (H3.2) *hold. Then $\bar{u}^\lambda(\cdot)$ converges in $L^2(s, T; \mathbb{R}^m)$ as $\lambda \to \infty$, denote the limit by $\bar{u}(\cdot)$. Moreover, there exists an operator $\Theta(t, s) \in \mathscr{L}(Z, \mathbb{R}^m)$ such that $\bar{u}(t) = \Theta(t, s)\mathbf{Z}_0$, and $\bar{u}(\cdot)$ is the optimal control of* Problem (P_0).

Proof In this proof we simply write $\langle \cdot, \cdot \rangle_{M^2}$, $\langle \cdot, \cdot \rangle_{L^2}$ as $\langle \cdot, \cdot \rangle$ if no ambiguity exists. By (3.2.3)(a), we have

$$\begin{aligned} &\langle P_0(s)\mathbf{Z}_0, \mathbf{Z}_0 \rangle \geqslant J_0^\lambda(s, \mathbf{Z}_0; \bar{u}^\lambda(\cdot)) = \langle \mathbf{G}\bar{\mathbf{Z}}^\lambda(T), \bar{\mathbf{Z}}^\lambda(T) \rangle \\ &+ \int_s^T \Big[\langle \mathbf{Q}(t)\bar{\mathbf{Z}}^\lambda(t), \bar{\mathbf{Z}}^\lambda(t) \rangle + \langle \mathbf{R}(t)\bar{u}^\lambda(t), \bar{u}^\lambda(t) \rangle \Big] dt, \end{aligned} \tag{3.2.12}$$

where for $z \in Z$,

$$P_0(t)z = \mathbf{T}(T, t)^*\mathbf{G}\mathbf{T}(T, t)z + \int_t^T \mathbf{T}(r, t)^*\mathbf{Q}(r)\mathbf{T}(r, t)zdr, \quad t \in [s, T].$$

Since $\mathbf{G} \geqslant 0$ and $\mathbf{Q} \geqslant 0$, $\{\bar{u}^\lambda(\cdot)\}$ is uniformly bounded in $L^2(s, T; \mathbb{R}^m)$ and hence there exists a subsequence, still denoted by $\{\bar{u}^\lambda(\cdot)\}$, converges weakly to $\bar{u}(\cdot)$ in $L^2(s, T; \mathbb{R}^m)$. By Lemma 3.2.4, $\bar{\mathbf{Y}}_t^\lambda = e^{D(t-s)}\psi + \int_s^t e^{D(t-r)}\mathcal{D}^\lambda \bar{u}^\lambda(r)dr$, converges weakly in L^2 to $\bar{\mathbf{Y}}_t = e^{D(t-s)}\psi + \int_s^t e^{D(t-r)}\mathcal{D}\bar{u}(r)dr$. Next we aim to prove that

$$\bar{\mathbf{X}}^\lambda(t) = \Phi(t, s)\xi + \int_s^t \Phi(t, r)\widetilde{B}(r)\bar{\mathbf{Y}}_r^\lambda dr,$$

also converges weakly to

$$\bar{\mathbf{X}}(t) = \Phi(t, s)\xi + \int_s^t \Phi(t, r)\widetilde{B}(r)\bar{\mathbf{Y}}_r dr.$$

Apparently it is sufficient to prove that for any $k \in M^2$,

$$\left\langle \int_s^t \Phi(t, r)\widetilde{B}(r)\bar{\mathbf{Y}}_r^\lambda dr, k \right\rangle_{M^2} \to \left\langle \int_s^t \Phi(t, r)\widetilde{B}(r)\bar{\mathbf{Y}}_r dr, k \right\rangle_{M^2}. \tag{3.2.13}$$

Noting for any $Y \in L^2$,

$$\begin{aligned}
&\left[(\lambda I - D)^{-1}Y\right](\theta) = \left[\textstyle\int_0^\infty e^{-\lambda t}e^{Dt}Y dt\right](\theta) \\
&= \int_0^{-\theta} e^{-\lambda t}Y(t+\theta)dt = \int_\theta^0 e^{\lambda(\theta-\beta)}Y(\beta)d\beta, \quad \theta \in [-\delta, 0],
\end{aligned}$$

it follows that for $t \in [s, T]$,

$$\begin{aligned}
&\left[\int_s^t e^{D(t-r)}\mathcal{D}^\lambda \bar{u}^\lambda(r)dr\right](\theta) \\
&= \left[-\int_s^t e^{D(t-r)}\lambda D(\lambda I - D)^{-1}F\bar{u}^\lambda(r)dr\right](\theta) \\
&= \left[\int_s^t e^{D(t-r)}\lambda e^{\lambda\cdot}\bar{u}^\lambda(r)dr\right](\theta) = \int_{(t+\theta)\vee s}^t \lambda\bar{u}^\lambda(r)e^{\lambda(t-r+\theta)}dr \\
&= \int_0^{(-\theta)\wedge(t-s)} \lambda\bar{u}^\lambda(t-\beta)e^{\lambda(\beta+\theta)}d\beta \\
&= \int_{\theta\vee(s-t)}^0 \lambda\bar{u}^\lambda(t+\beta)e^{\lambda(\theta-\beta)}d\beta, \quad \theta \in [-\delta, 0].
\end{aligned}$$

Hence we deduce

$$\int_s^t e^{D(t-r)}\mathcal{D}^\lambda \bar{u}^\lambda(r)dr = \int_{(s-t)\vee\cdot}^0 \lambda\bar{u}^\lambda(t+\beta)e^{\lambda(\cdot-\beta)}d\beta.$$

Similar to (2.1.16), we have

$$\int_s^t \Phi(t,r)\begin{pmatrix}\int_{[-\delta,0]} B(r,\theta)\bar{\mathbf{Y}}_r^\lambda(\theta)\mu(d\theta)\\ 0\end{pmatrix}dr$$
$$=\begin{pmatrix}\int_s^t\int_{[-\delta,0]}\Phi^0(t,r)B(r,\theta)\bar{\mathbf{Y}}_r^\lambda(\theta)\mu(d\theta)dr\\ \displaystyle\int_s^t\int_{[-\delta,0]}\Phi^0(t+\cdot,r)B(r,\theta)\bar{\mathbf{Y}}_r^\lambda(\theta)\mu(d\theta)dr\end{pmatrix}$$
$$=\begin{pmatrix}\int_{[-\delta,0]}\int_{s+\theta}^{t+\theta}\Phi^0(t,r-\theta)B(r-\theta,\theta)\bar{\mathbf{Y}}_{r-\theta}^\lambda(\theta)dr\mu(d\theta)\\ \displaystyle\int_{[-\delta,0]}\int_{s+\theta}^{t+\theta}\Phi^0(t+\cdot,r-\theta)B(r-\theta,\theta)\bar{\mathbf{Y}}_{r-\theta}^\lambda(\theta)dr\mu(d\theta)\end{pmatrix},$$

which implies that

$$\begin{aligned}
&\Big\langle k,\int_s^t \Phi(t,r)\begin{pmatrix}\int_{[-\delta,0]} B(r,\theta)\bar{\mathbf{Y}}_r^\lambda(\theta)\mu(d\theta)\\ 0\end{pmatrix}dr\Big\rangle\\
&=\Big\langle k^0,\int_{[-\delta,0]}\int_{s+\theta}^{t+\theta}\Phi^0(t,r-\theta)B(r-\theta,\theta)\bar{\mathbf{Y}}_{r-\theta}^\lambda(\theta)dr\mu(d\theta)\Big\rangle\\
&\quad+\int_{-\delta}^0\Big\langle k^1(\alpha),\int_{[-\delta,0]}\int_{s+\theta}^{t+\theta}\Phi^0(t+\alpha,r-\theta)B(r-\theta,\theta)\bar{\mathbf{Y}}_{r-\theta}^\lambda(\theta)dr\mu(d\theta)\Big\rangle d\alpha\\
&=\Big\langle k^0,\int_{[-\delta,0]}\int_{s+\theta}^{t+\theta}\Phi^0(t,r-\theta)B(r-\theta,\theta)\Big(e^{D(r-\theta-s)}\psi\\
&\quad+\int_s^{r-\theta}e^{D(r-\theta-\theta')}\mathcal{D}^\lambda\bar{u}^\lambda(\theta')d\theta'\Big)(\theta)dr\mu(d\theta)\Big\rangle\\
&\quad+\int_{-\delta}^0\Big\langle k^1(\alpha),\int_{[-\delta,0]}\int_{s+\theta}^{t+\theta}\Phi^0(t+\alpha,r-\theta)B(r-\theta,\theta)\Big(e^{D(r-\theta-s)}\psi\\
&\quad+\int_s^{r-\theta}e^{D(r-\theta-\theta')}\mathcal{D}^\lambda\bar{u}^\lambda(\theta')d\theta'\Big)(\theta)dr\mu(d\theta)\Big\rangle d\alpha\\
&=\Big\langle k^0,\int_{[-\delta,0]}\int_s^t\Phi^0(t,r)B(r,\theta)\psi(r-s+\theta)\mathbf{1}_{[-\delta,s-r]}(\theta)dr\mu(d\theta)\Big\rangle\\
&\quad+\int_{-\delta}^0\Big\langle k^1(\alpha),\int_{[-\delta,0]}\int_s^t\Phi^0(t+\alpha,r)B(r,\theta)\psi(r-s\\
&\quad+\theta)\mathbf{1}_{[-\delta,s-r]}(\theta)dr\mu(d\theta)\Big\rangle d\alpha\\
&\quad+\Big\langle k^0,\int_{[-\delta,0]}\int_s^t\Phi^0(t,r)B(r,\theta)\int_{(s-r)\vee\theta}^0\lambda\bar{u}^\lambda(r+\beta)e^{\lambda(\theta-\beta)}d\beta dr\mu(d\theta)\Big\rangle\\
&\quad+\int_{-\delta}^0\Big\langle k^1(\alpha),\int_{[-\delta,0]}\int_s^t\Phi^0(t+\alpha,r)B(r,\theta)\\
&\quad\times\int_{(s-r)\vee\theta}^0\lambda\bar{u}^\lambda(r+\beta)e^{\lambda(\theta-\beta)}d\beta dr\mu(d\theta)\Big\rangle d\alpha.
\end{aligned}$$

By (2.2.14), it follows that

$$
\begin{aligned}
&\Big\langle k, \int_s^t \Phi(t,r)\begin{pmatrix}\int_{[-\delta,0]} B(r,\theta)\bar{\mathbf{Y}}_r^{\lambda}(\theta)\mu(d\theta)\\ 0\end{pmatrix} dr\Big\rangle \\
&= \int_{[-\delta,0]}\int_s^t \Big\langle [\Phi(t,r)^*k]^0, B(r,\theta)\Big[\psi(r-s+\theta)\mathbf{1}_{[-\delta,s-r]}(\theta) \\
&\qquad + \int_{(s-r)\vee\theta}^0 \lambda\bar{u}^{\lambda}(r+\beta)e^{\lambda(\theta-\beta)}d\beta\Big]\Big\rangle dr\,\mu(d\theta),
\end{aligned}
$$

where $[\Phi(t,r)^*k]^0 \in \mathbb{R}^n$ is the component of $\Phi(t,r)^*k$. Similarly, by Lemma 2.1.4 we have

$$
\begin{aligned}
&\Big\langle k, \int_s^t \Phi(t,r)\begin{pmatrix}\int_{[-\delta,0]} B(r,\theta)\bar{\mathbf{Y}}_r(\theta)\mu(d\theta)\\ 0\end{pmatrix} dr\Big\rangle \\
&= \int_{[-\delta,0]}\int_s^t \langle [\Phi(t,r)^*k]^0, B(r,\theta)\Big[\psi(r-s+\theta)\mathbf{1}_{[-\delta,s-r]}(\theta) \\
&\qquad + \bar{u}(r+\theta)\mathbf{1}_{(s-r,0]}(\theta)\Big]\rangle dr\,\mu(d\theta).
\end{aligned}
$$

To prove (3.2.13), next we show that

$$
\begin{aligned}
&\int_{[-\delta,0]}\int_s^t \langle [\Phi(t,r)^*k]^0, B(r,\theta)\int_{(s-r)\vee\theta}^0 \lambda\bar{u}^{\lambda}(r+\beta)e^{\lambda(\theta-\beta)}d\beta\rangle dr\,\mu(d\theta) \\
&\to \int_{[-\delta,0]}\int_s^t \langle [\Phi(t,r)^*k]^0, B(r,\theta)\bar{u}(r+\theta)\mathbf{1}_{(s-r,0]}(\theta)\rangle dr\,\mu(d\theta), \text{ as } \lambda\to\infty.
\end{aligned}
\tag{3.2.14}
$$

Without loss of generality, assume $t > s+\delta$. By the boundedness of $\Phi(\cdot,\cdot)^*$ and $B(\cdot,\cdot)$, we have

$$
\begin{aligned}
&\Big|\int_{[-\delta,0]}\int_s^t \langle [\Phi(t,r)^*k]^0, B(r,\theta)\int_{(s-r)\vee\theta}^0 \lambda\bar{u}^{\lambda}(r+\beta)e^{\lambda(\theta-\beta)}d\beta\rangle dr\,\mu(d\theta) \\
&\quad - \int_{[-\delta,0]}\int_s^t \langle [\Phi(t,r)^*k]^0, B(r,\theta)\bar{u}(r+\theta)\mathbf{1}_{(s-r,0]}(\theta)\rangle dr\,\mu(d\theta)\Big| \\
&= \Big|\int_{[-\delta,0]}\int_s^{(s-\theta)\wedge t} \langle [\Phi(t,r)^*k]^0, B(r,\theta)\int_{s-r}^0 \lambda\bar{u}^{\lambda}(r+\beta)e^{\lambda(\theta-\beta)}d\beta\rangle dr\,\mu(d\theta) \\
&\quad + \int_{[-\delta,0]}\int_{(s-\theta)\wedge t}^t \langle [\Phi(t,r)^*k]^0, B(r,\theta)\int_{\theta}^0 \lambda\bar{u}^{\lambda}(r+\beta)e^{\lambda(\theta-\beta)}d\beta\rangle dr\,\mu(d\theta) \\
&\quad - \int_{[-\delta,0]}\int_s^t \langle [\Phi(t,r)^*k]^0, B(r,\theta)\bar{u}(r+\theta)\mathbf{1}_{(s-r,0]}(\theta)\rangle dr\,\mu(d\theta)\Big|
\end{aligned}
$$

$$
\begin{aligned}
&\leqslant K\int_{[-\delta,0]}\int_s^{(s-\theta)\wedge t}\int_{s-r}^0 \lambda|\bar u^\lambda(r+\beta)|e^{\lambda(\theta-\beta)}d\beta dr\mu(d\theta)\\
&\quad+\Big|\int_{[-\delta,0]}\int_{(s-\theta)\wedge t}^t \big\langle[\Phi(t,r)^*k]^0,\\
&\qquad B(r,\theta)\int_\theta^0 \lambda\bar u^\lambda(r+\beta)e^{\lambda(\theta-\beta)}d\beta\big\rangle dr\mu(d\theta)\\
&\quad-\int_{[-\delta,0]}\int_{s-\theta}^t \big\langle[\Phi(t,r)^*k]^0, B(r,\theta)\bar u(r+\theta)\big\rangle dr\mu(d\theta)\Big|\\
&= K\int_{[-\delta,0]}\int_{\theta\vee(s-t)}^0\int_s^{(s-\theta+\beta)\wedge(t+\beta)} \lambda|\bar u^\lambda(r)|e^{\lambda(\theta-\beta)}drd\beta\mu(d\theta)\\
&\quad+\Big|\int_{[-\delta,0]}\int_{(s-\theta)\wedge t}^t \big\langle[\Phi(t,r)^*k]^0,\\
&\qquad B(r,\theta)\int_\theta^0 \lambda\bar u^\lambda(r+\beta)e^{\lambda(\theta-\beta)}d\beta\big\rangle dr\mu(d\theta)\\
&\quad-\int_{[-\delta,0]}\int_{s-\theta}^t \big\langle[\Phi(t,r)^*k]^0, B(r,\theta)\bar u(r+\theta)\big\rangle dr\mu(d\theta)\Big|\\
&:= I_1+I_2,
\end{aligned}
$$

in this proof K is a constant changing from line to line. Since $\{\bar u^\lambda(\cdot)\}$ is uniformly bounded in $L^2(s,T;\mathbb{R}^n)$ and $\int_s^{(s-\theta+\beta)\wedge(t+\beta)}|\bar u^\lambda(r)|dr$ is continuous at $\beta=\theta\vee(s-t)$, for any $\varepsilon>0$, there exists α_θ such that when $\beta\in[\theta\vee(s-t),\alpha_\theta]$,

$$
\int_s^{(s-\theta+\beta)\wedge(t+\beta)}|\bar u^\lambda(r)|dr<\varepsilon,
$$

which yields

$$
\begin{aligned}
I_1 &\leqslant K\int_{[-\delta,0]}\Big(\int_{\theta\vee(s-t)}^{\alpha_\theta}+\int_{\alpha_\theta}^0\Big)\int_s^{(s-\theta+\beta)\wedge(t+\beta)}|\bar u^\lambda(r)|dr\lambda e^{\lambda(\theta-\beta)}d\beta\mu(d\theta)\\
&\leqslant K\varepsilon\int_{[-\delta,0]}\int_{\theta\vee(s-t)}^{\alpha_\theta}\lambda e^{\lambda(\theta-\beta)}d\beta\mu(d\theta)\\
&\quad+K\int_{[-\delta,0]}\int_{\alpha_\theta}^0\int_s^{(s-\theta+\beta)\wedge(t+\beta)}\lambda|\bar u^\lambda(r)|e^{\lambda(\theta-\beta)}drd\beta\mu(d\theta)\\
&= K\varepsilon\int_{[-\delta,0]}[e^{\lambda(0\wedge(\theta-s+t))}-e^{\lambda(\theta-\alpha_\theta)}]\mu(d\theta)\\
&\quad+K\int_{[-\delta,0]}\int_{\alpha_\theta}^0\int_s^{(s-\theta+\beta)\wedge(t+\beta)}\lambda|\bar u^\lambda(r)|e^{\lambda(\theta-\beta)}drd\beta\mu(d\theta),
\end{aligned}
$$

it is clear that

$$K\varepsilon \int_{[-\delta,0]} [e^{\lambda(0\wedge(\theta-s+t))} - e^{\lambda(\theta-\alpha_\theta)}]\mu(d\theta) \leqslant K\varepsilon,$$

and by the uniform boundedness of $||\bar{u}^\lambda(\cdot)||$, we get

$$K\int_{[-\delta,0]}\int_{\alpha_\theta}^{0}\int_{s}^{(s-\theta+\beta)\wedge(t+\beta)} \lambda|\bar{u}^\lambda(r)|e^{\lambda(\theta-\beta)}drd\beta\mu(d\theta) \to 0, \text{ as } \lambda \to 0,$$

thus by the arbitrariness of ε, we obtain

$$I_1 \to 0, \text{ as } \lambda \to 0. \tag{3.2.15}$$

Noting

$$\begin{aligned}
I_2 \leqslant & \int_{[-\delta,0]} \Big| \int_{\theta}^{\alpha'_\theta}\int_{s-\theta+\beta}^{t+\beta} \langle [\Phi(t,r-\beta)^*k]^0, B(r-\beta,\theta)\bar{u}^\lambda(r)\rangle dr\lambda e^{\lambda(\theta-\beta)}d\beta \\
& - \int_{s-\theta}^{t} \langle [\Phi(t,r)^*k]^0, B(r,\theta)\bar{u}(r+\theta)\rangle dr \Big| \mu(d\theta) \\
& + \int_{[-\delta,0]} \Big| \int_{\alpha'_\theta}^{0}\int_{s-\theta+\beta}^{t+\beta} \langle [\Phi(t,r-\beta)^*k]^0, B(r-\beta,\theta)\bar{u}^\lambda(r)\rangle dr \\
& \qquad \lambda e^{\lambda(\theta-\beta)}d\beta \Big| \mu(d\theta) \\
\leqslant & \int_{[-\delta,0]} \Big| \int_{\theta}^{\alpha'_\theta}\int_{s-\theta+\beta}^{t+\beta} \langle [\Phi(t,r-\beta)^*k]^0, B(r-\beta,\theta)\bar{u}^\lambda(r)\rangle dr\lambda e^{\lambda(\theta-\beta)}d\beta \\
& - \int_{\theta}^{\alpha'_\theta}\int_{s}^{t+\theta} \langle [\Phi(t,r-\theta)^*k]^0, B(r-\theta,\theta)\bar{u}(r)\rangle dr \frac{1}{\alpha'_\theta-\theta}d\beta \Big| \mu(d\theta) \\
& + \int_{[-\delta,0]} \Big| \int_{\alpha'_\theta}^{0}\int_{s-\theta+\beta}^{t+\beta} \langle [\Phi(t,r-\beta)^*k]^0, B(r-\beta,\theta)\bar{u}^\lambda(r)\rangle dr \\
& \qquad \lambda e^{\lambda(\theta-\beta)}d\beta \Big| \mu(d\theta) \\
\leqslant & \int_{[-\delta,0]}\int_{\theta}^{\alpha'_\theta}\int_{s-\theta+\beta}^{t+\beta} \Big| B(r-\beta,\theta)^\top[\Phi(t,r-\beta)^*k]^0 \\
& \qquad - B(r-\theta,\theta)^\top[\Phi(t,r-\theta)^*k]^0 \Big| |\bar{u}^\lambda(r)|dr\lambda e^{\lambda(\theta-\beta)}d\beta\mu(d\theta) \\
& + \int_{[-\delta,0]} \Big| \int_{\theta}^{\alpha'_\theta}\int_{s-\theta+\beta}^{t+\beta} \langle B(r-\theta,\theta)^\top[\Phi(t,r-\theta)^*k]^0, \bar{u}^\lambda(r)\rangle dr\lambda e^{\lambda(\theta-\beta)}d\beta \\
& - \int_{\theta}^{\alpha'_\theta}\int_{s}^{t+\theta} \langle B(r-\theta,\theta)^\top[\Phi(t,r-\theta)^*k]^0, \bar{u}(r)\rangle dr \frac{1}{\alpha'_\theta-\theta}d\beta \Big| \mu(d\theta)
\end{aligned}$$

$$+\int_{[-\delta,0]}\Big|\int_{\alpha'_\theta}^{0}\int_{s-\theta+\beta}^{t+\beta}\big\langle[\Phi(t,r-\beta)^*k]^0, B(r-\beta,\theta)\bar{u}^\lambda(r)\big\rangle dr \times \lambda e^{\lambda(\theta-\beta)}d\beta\Big|\mu(d\theta)$$

$$:= I_2^1 + I_2^2 + I_2^3. \tag{3.2.16}$$

By the boundedness of $B(\cdot,\cdot)$ and $\Phi(\cdot,\cdot)^*$, for any $\varepsilon > 0$, we can choose α''_θ such that

$$\int_{\theta\vee(r-t)}^{\alpha''_\theta\wedge(r+\theta-s)}|B(r-\beta,\theta)^\top[\Phi(t,r-\beta)^*k]^0 - B(r-\theta,\theta)^\top[\Phi(t,r-\theta)^*k]^0|d\beta < \varepsilon,$$

which implies that

$$\begin{aligned} I_2^1 \leqslant \int_{[-\delta,0]}\int_s^{t+\alpha''_\theta}\int_{\theta\vee(r-t)}^{\alpha''_\theta\wedge(r+\theta-s)}&|B(r-\beta,\theta)^\top[\Phi(t,r-\beta)^*k]^0 \\ &-B(r-\theta,\theta)^\top[\Phi(t,r-\theta)^*k]^0|d\beta\lambda|\bar{u}^\lambda(r)|dr\mu(d\theta) \leqslant K\lambda\varepsilon. \end{aligned} \tag{3.2.17}$$

Noting $\int_{s-\theta+\beta}^{t+\beta}\big\langle B(r-\theta,\theta)^\top[\Phi(t,r-\theta)^*x]^0, \bar{u}^\lambda(r)\big\rangle dr$ is continuous at $\beta=\theta$, for any $\varepsilon > 0$, there exists α'''_θ such that when $\beta \in [\theta, \alpha'''_\theta]$,

$$\begin{aligned} \Big|&\int_{s-\theta+\beta}^{t+\beta}\big\langle B(r-\theta,\theta)^\top[\Phi(t,r-\theta)^*x]^0, \bar{u}^\lambda(r)\big\rangle dr \\ &-\int_s^{t+\theta}\big\langle B(r-\theta,\theta)^\top[\Phi(t,r-\theta)^*x]^0, \bar{u}^\lambda(r)\big\rangle dr\Big| < \varepsilon, \end{aligned}$$

which yields

$$\begin{aligned} I_2^2 \leqslant \int_{[-\delta,0]}\int_\theta^{\alpha'''_\theta}\Big|&\int_{s-\theta+\beta}^{t+\beta}\big\langle B(r-\theta,\theta)^\top[\Phi(t,r-\theta)^*k]^0, \bar{u}^\lambda(r)\big\rangle dr \\ &-\int_s^{t+\theta}\big\langle B(r-\theta,\theta)^\top[\Phi(t,r-\theta)^*k]^0, \bar{u}^\lambda(r)\big\rangle dr\Big|\lambda e^{\lambda(\theta-\beta)}d\beta\mu(d\theta) \\ &+\int_{[-\delta,0]}\Big|\int_\theta^{\alpha'''_\theta}\int_s^{t+\theta}\big\langle B(r-\theta,\theta)^\top[\Phi(t,r-\theta)^*k]^0, \bar{u}^\lambda(r)\big\rangle dr \end{aligned}$$

$$
\begin{aligned}
&\times(\lambda e^{\lambda(\theta-\beta)} - \frac{1}{\alpha_\theta''' - \theta})d\beta\Big|\mu(d\theta)\\
&+\int_{[-\delta,0]}\Big|\int_s^{t+\theta}\big\langle B(r-\theta,\theta)^\top[\Phi(t,r-\theta)^*k]^0, \bar{u}^\lambda(r)-\bar{u}(r)\big\rangle dr\Big|\mu(d\theta)\\
\leqslant &\int_{[-\delta,0]}\varepsilon[1-e^{\lambda(\theta-\alpha_\theta''')}]\mu(d\theta)\\
&+K\int_{[-\delta,0]}\int_s^{t+\theta}|\bar{u}^\lambda(r)|dr e^{\lambda(\theta-\alpha_\theta''')}\mu(d\theta)\\
&+\int_{[-\delta,0]}\Big|\int_s^{t+\theta}\big\langle B(r-\theta,\theta)^\top[\Phi(t,r-\theta)^*k]^0, \bar{u}^\lambda(r)-\bar{u}(r)\big\rangle dr\Big|\mu(d\theta).
\end{aligned}
\tag{3.2.18}
$$

We can choose $\alpha_\theta' = \alpha_\theta'' \wedge \alpha_\theta'''$, recalling $\bar{u}^\lambda(\cdot)$ converges weakly to $\bar{u}(\cdot)$ in $L^2(s,T;\mathbb{R}^m)$, then by (3.2.16), (3.2.17), (3.2.18), the arbitrariness of ε and the uniform boundedness of $||\bar{u}^\lambda(\cdot)||$, we deduce

$$
I_2 \to 0, \text{ as } \lambda \to 0,
$$

which and (3.2.15) imply (3.2.14), thus (3.2.13) holds. Therefore

$$
\bar{\mathbf{Z}}^\lambda(t) = \mathbf{T}(t,s)\mathbf{Z}_0 + \int_s^t \mathbf{T}(t,r)\mathbf{B}^\lambda\bar{u}^\lambda(r)dr,
$$

converges weakly to

$$
\bar{\mathbf{Z}}(t) = \mathbf{T}(t,s)\mathbf{Z}_0 + \int_s^t \mathbf{T}(t,r)\mathbf{B}\bar{u}(r)dr.
$$

It follows that

$$
J_0(s,\mathbf{Z}_0;\bar{u}(\cdot)) = \lim_{\lambda\to\infty} J_0^\lambda(s,\mathbf{Z}_0;\bar{u}^\lambda(\cdot)),
$$

which implies that the $L^2(s,T;\mathbb{R}^m)$ norm of $\bar{u}^\lambda(\cdot)$ converges to that of $\bar{u}(\cdot)$, thus $\bar{u}^\lambda(\cdot)$ converges strongly in $L^2(s,T;\mathbb{R}^m)$ to $\bar{u}(\cdot)$. Noting

$$
\begin{aligned}
J_0(s,\mathbf{Z}_0;\bar{u}(\cdot)) &= \lim_{\lambda\to\infty} J_0^\lambda(s,\mathbf{Z}_0;\bar{u}^\lambda(\cdot))\\
&\leqslant \lim_{\lambda\to\infty} J_0^\lambda(s,\mathbf{Z}_0;u(\cdot)) = J_0(s,\mathbf{Z}_0;u(\cdot)), \quad \forall\, u(\cdot)\in L^2(s,T;\mathbb{R}^m),
\end{aligned}
$$

hence we conclude that $\bar{u}(\cdot)$ is the optimal control of Problem (P_0). Next we show that the original sequence $\{\bar{u}^{\lambda}(\cdot)\}$ also converges strongly to $\bar{u}(\cdot)$. Suppose there are a subsequence $\{\bar{u}^{\lambda_n}(\cdot)\}$ does not conveges to $\bar{u}(\cdot)$, then we can extract a subsequence $\{\bar{u}^{\lambda_{n_k}}\}$, which lies outside of a neighborhood of $\bar{u}(\cdot)$ in $L^2(s, T; \mathbb{R}^m)$, converges weakly in $L^2(s, T; \mathbb{R}^m)$ to some element $\tilde{u}(\cdot)$, then repeating the above argument we also conclude that $\tilde{u}(\cdot)$ is the optimal control of Problem (P_0). By strictly convexity of $J_0(s, \mathbf{Z}_0; \cdot)$, we have $\bar{u}(\cdot) = \tilde{u}(\cdot)$, which is a contradiction. Finally, by (3.2.11), we get

$$\bar{u}^{\lambda}(t) = \Theta^{\lambda}(t)\mathbf{T}^{\lambda}_{\Theta}(t, s)\mathbf{Z}_0, \quad \text{a.e. } t \in [s, T],$$

hence there exists an operator $\Theta(t, s) \in \mathscr{L}(Z, \mathbb{R}^m)$ such that $\bar{u}(t) = \Theta(t, s)\mathbf{Z}_0$. This completes the proof of Lemma 3.2.6. □

Lemma 3.2.7 *Let* (H3.2) *hold. Then for any* $t \in [s, T]$, $\mathbf{T}^{\lambda}_{\Theta}(t, s)$, *satisfying (3.2.4), converges strongly to some operator* $\mathbf{T}_{\Theta}(t, s)$, *and* $\mathbf{T}_{\Theta}(t, s)$ *satisfies the following integral equation:*

$$\mathbf{T}_{\Theta}(t, s)\mathbf{Z}_0 = \mathbf{T}(t, s)\mathbf{Z}_0 + \int_s^t \mathbf{T}(t, r)\mathbf{B}\Theta(r, s)\mathbf{Z}_0 dr, \tag{3.2.19}$$

and there is a constant $K > 0$, *independent of* t, s, *such that* $\|\mathbf{T}_{\Theta}(t, s)\|_{\mathscr{L}(Z)} \leqslant K$.

Proof First we show that there exists a constant $K > 0$, independent of λ, such that

$$\begin{aligned} &\sup_{s \leqslant t \leqslant T} \left\| \int_s^t \mathbf{T}(t, r)\mathbf{B}^{\lambda} u(r) dr \right\|_Z \\ &\quad \leqslant K \|u(\cdot)\|_{L^2(s,T;\mathbb{R}^m)}, \quad \forall\, u(\cdot) \in L^2(s, T; \mathbb{R}^m). \end{aligned} \tag{3.2.20}$$

Noting

$$\left[(\lambda I - D)^{-1} F u(r)\right](\theta) = \int_0^{-\theta} e^{-\lambda t} dt u(r) = \frac{1}{\lambda}\left[1 - e^{\lambda\theta}\right] u(r), \quad \theta \in [-\delta, 0],$$

we have

$$\begin{aligned} & e^{D(t-r)}\left[(\lambda I - D)^{-1} F u(r)\right](\theta) \\ &= \begin{cases} \begin{cases} \frac{1}{\lambda}\left[1 - e^{\lambda(t+\theta-r)}\right] u(r), & -\delta \leqslant \theta \leqslant r - t, \\ 0, & r - t < \theta \leqslant 0, \end{cases} & \text{if } t - r \leqslant \delta, \\ 0, \qquad\qquad\qquad -\delta \leqslant \theta \leqslant 0, & \text{if } t - r > \delta. \end{cases} \end{aligned}$$

It follows that

$$
\begin{aligned}
&-\Big[D\int_s^t e^{D(t-r)}(\lambda I-D)^{-1}Fu(r)dr\Big](\theta)\\
&=\int_{(t+\theta)\vee s}^t e^{\lambda(t+\theta-r)}u(r)dr,\quad -\delta\leqslant\theta\leqslant 0,
\end{aligned}
$$

which implies that

$$
\begin{aligned}
&\Big[(\lambda I-D)\int_s^t e^{D(t-r)}(\lambda I-D)^{-1}Fu(r)dr\Big](\theta)\\
&=\int_{(t+\theta)\vee s}^t u(r)dr=\int_s^t e^{D(t-r)}Fu(r)dr(\theta),\qquad \theta\in[-\delta,0].
\end{aligned}
$$

Hence we derive

$$
\int_s^t(\lambda I-D)^{-1}e^{D(t-r)}Fu(r)dr=\int_s^t e^{D(t-r)}(\lambda I-D)^{-1}Fu(r)dr,
$$

which yields

$$
\begin{aligned}
&\int_s^t e^{D(t-r)}\mathcal{D}^\lambda u(r)dr=-\int_s^t e^{D(t-r)}\lambda D(\lambda I-D)^{-1}Fu(r)dr\\
&=\lambda\int_s^t e^{D(t-r)}(\lambda I-D)(\lambda I-D)^{-1}Fu(r)dr\\
&\quad-\lambda^2\int_s^t e^{D(t-r)}(\lambda I-D)^{-1}Fu(r)dr\\
&=\int_s^t\Big[\lambda e^{D(t-r)}Fu(r)-\lambda^2(\lambda I-D)^{-1}e^{D(t-r)}Fu(r)\Big]dr\\
&=-\lambda\int_s^t(\lambda I-D)^{-1}De^{D(t-r)}Fu(r)dr.
\end{aligned}
\tag{3.2.21}
$$

Since $\lambda(\lambda I-D)^{-1}Y\to Y$ as $\lambda\to\infty$ for any $Y\in L^2$, we have

$$
\begin{aligned}
&\Big\|\int_s^t e^{D(t-r)}\mathcal{D}^\lambda u(r)dr\Big\|_{L^2}=\Big\|\lambda\int_s^t(\lambda I-D)^{-1}De^{D(t-r)}Fu(r)dr\Big\|_{L^2}\\
&\leqslant\Big\|D\int_s^t e^{D(t-r)}Fu(r)dr\Big\|_{L^2}+\varepsilon.
\end{aligned}
\tag{3.2.22}
$$

Noting

$$\Big[D\int_s^t e^{D(t-r)}Fu(r)dr\Big](\theta)=\begin{cases}-u(t+\theta), & s-t\leqslant\theta\leqslant 0,\\ 0, & -\delta\leqslant\theta<s-t,\end{cases}$$

we deduce

$$\Big\|D\int_s^t e^{D(t-r)}Fu(r)dr\Big\|_{L^2}^2\leqslant\|u(\cdot)\|_{L^2(s,T;\mathbb{R}^m)}. \tag{3.2.23}$$

By the arbitrariness of ε, (3.2.22) and (3.2.23) yield

$$\Big\|\int_s^t e^{D(t-r)}\mathcal{D}^\lambda u(r)dr\Big\|_{L^2}\leqslant\|u(\cdot)\|_{L^2(s,T;\mathbb{R}^m)}. \tag{3.2.24}$$

Then by the boundedness of $\Phi^0(\cdot,\cdot)$ and $B(\cdot,\cdot)$, we have

$$\begin{aligned}&\Big\|\int_s^t\int_r^t\Phi(t,\beta)\widetilde{B}(\beta)e^{D(\beta-r)}\mathcal{D}^\lambda u(r)d\beta dr\Big\|_{M^2}\\ &\leqslant K\int_s^t\int_{[-\delta,0]}\int_r^{t\wedge(r-\theta)}\lambda e^{\lambda(\beta-r+\theta)}|u(r)|d\beta\mu(d\theta)dr\\ &\leqslant K\|u(\cdot)\|_{L^2(s,T;\mathbb{R}^m)},\quad\forall\,u(\cdot)\in L^2(s,T;\mathbb{R}^m),\end{aligned}$$

which and (3.2.24) imply (3.2.20). Here $K>0$ is a constant and will change from line to line in our proof. Recall $\bar{u}^\lambda(\cdot)\to\bar{u}(\cdot)$ in $L^2(s,T;\mathbb{R}^m)$, then we have

$$\Big\|\int_s^t\mathbf{T}(t,r)\mathbf{B}^\lambda(\bar{u}^\lambda(r)-\bar{u}(r))dr\Big\|_Z\to 0,\qquad\text{as }\lambda\to 0. \tag{3.2.25}$$

Next we aim to prove that

$$\Big\|\int_s^t\mathbf{T}(t,r)\big[\mathbf{B}^\lambda-\mathbf{B}\big]\bar{u}(r)dr\Big\|_Z\to 0,\qquad\text{as }\lambda\to 0. \tag{3.2.26}$$

By (3.2.21) and Proposition 1.3.4, we have

$$\begin{aligned}&\Big\|\int_s^t e^{D(t-r)}\mathcal{D}^\lambda\bar{u}(r)dr-\int_s^t e^{D(t-r)}\mathcal{D}\bar{u}(r)dr\Big\|_{L^2}\\ =&\Big\|\big(D-\lambda D(\lambda I-D)^{-1}\big)\int_s^t e^{D(t-r)}F\bar{u}(r)dr\Big\|_{L^2}\to 0,\qquad\text{as }\lambda\to 0,\end{aligned}$$

which yields

$$\int_s^t e^{D(t-r)}\mathcal{D}^\lambda\bar{u}(r)dr\to\int_s^t e^{D(t-r)}\mathcal{D}\bar{u}(r)dr\text{ in }L^2,\qquad\text{as }\lambda\to\infty. \tag{3.2.27}$$

Using the similar proof in Lemma 3.2.6, we obtain

$$\int_s^t \int_r^t \Phi(t,\beta)\widetilde{B}(\beta)e^{D(\beta-r)}\mathcal{D}^\lambda \bar{u}(r)d\beta dr$$
$$\to \int_s^t \int_r^t \Phi(t,\beta)\widetilde{B}(\beta)e^{D(\beta-r)}\mathcal{D}\bar{u}(r)d\beta dr \quad \text{in } M^2,$$

which and (3.2.27) imply (3.2.26). Then by (3.2.25) and (3.2.26), we get

$$\int_s^t \mathbf{T}(t,r)\mathbf{B}^\lambda \bar{u}^\lambda(r)dr \to \int_s^t \mathbf{T}(t,r)\mathbf{B}\bar{u}(r)dr \text{ in } Z, \quad \text{as } \lambda \to \infty. \tag{3.2.28}$$

It follows that $\mathbf{T}^\lambda_\Theta(t,s)$, satisfying (3.2.4), strongly converges to the following integral equation:

$$\mathbf{T}_\Theta(t,s)\mathbf{Z}_0 = \mathbf{T}(t,s)\mathbf{Z}_0 + \int_s^t \mathbf{T}(t,r)\mathbf{B}\Theta(r,s)\mathbf{Z}_0 dr, \quad t \in [s,T].$$

Recalling (3.2.12), there exists a constant $K > 0$ such that $\|\bar{u}^\lambda(\cdot)\|_{L^2(s,T;\mathbb{R}^m)} \leqslant K\|\mathbf{Z}_0\|_Z$, and by (3.2.4), we have

$$\begin{aligned}\mathbf{T}^\lambda_\Theta(t,s)\mathbf{Z}_0 &= \mathbf{T}(t,s)\mathbf{Z}_0 - \int_s^t \mathbf{T}(t,r)\mathbf{B}^\lambda \mathbf{R}(r)^{-1}\big[\mathbf{B}^{\lambda *}P^\lambda(r) \\ &\quad +\mathbf{S}(r)\big]\mathbf{T}^\lambda_\Theta(r,s)\mathbf{Z}_0 dr \\ &= \mathbf{T}(t,s)\mathbf{Z}_0 + \int_s^t \mathbf{T}(t,r)\mathbf{B}^\lambda \bar{u}^\lambda(r)dr, \quad t \in [s,T].\end{aligned}$$

Then by (3.2.20), we obtain $\|\mathbf{T}^\lambda_\Theta(t,s)\mathbf{Z}_0\|_Z \leqslant K\|\mathbf{Z}_0\|_Z$, where K independent of λ, t, s, is a generic constant, thus $\|\mathbf{T}_\Theta(t,s)\|_{\mathscr{L}(Z)} \leqslant K$. This completes the proof of Lemma 3.2.7. □

Lemma 3.2.8 *Let* (H3.2) *hold. Denote* $\mathbf{B}^* := (0, \mathcal{D}^*)$ *(where* $\mathcal{D}^*$ *is understood in the generalized sense, see Lemma 3.2.5). Assume* μ *is absolutely continuous in lebesgue measure, then* $P^\lambda(t)$ *satisfying (3.2.3)(c) strongly converges to* $P(t)$*, which satisfies the following integral equation:*

$$P(t)z = \mathbf{T}(T,t)^*\mathbf{G}\mathbf{T}_\Theta(T,t)z + \int_t^T \mathbf{T}(r,t)^*\mathbf{Q}(r)\mathbf{T}_\Theta(r,t)z dr, \ t \in [s,T],$$

where $\mathbf{T}_\Theta(t,s)$ *is defined by (3.2.19). Moreover* $\mathbf{B}^*P(t) \in \mathscr{L}(Z,\mathbb{R}^m)$ *is well-defined,* $\mathbf{B}^{\lambda *}P^\lambda(t)z \to \mathbf{B}^*P(t)z$ *for all* $t \in [s,T]$*,* $z \in Z$ *and* $\sup_{s\leqslant t\leqslant T} \|\mathbf{B}^*P(t)\|_{\mathscr{L}(Z,\mathbb{R}^m)} < \infty$.

Proof It is obvious that $P^\lambda(t)$ strongly converges to $P(t)$ by Lemma 3.2.7. Next we show that $\mathbf{B}^*P(t)$ is well-defined. Recall (2.2.16), we have

$$\mathbf{T}(t,s)^*z = \begin{pmatrix} [\mathbf{T}(t,s)^*z]^0 \\ [\mathbf{T}(t,s)^*z]^1 \end{pmatrix}, \quad \forall z = \begin{pmatrix} \xi \\ \psi \end{pmatrix} \in Z,$$

where

$$\begin{aligned}
&[\mathbf{T}(t,s)^*z]^0 = \Phi(t,s)^*\xi, \\
&([\mathbf{T}(t,s)^*z]^1)(\theta) = \int_{[s+\theta-t,\theta]} B(s+\theta-\beta,\beta)^\top[\Phi(t,s+\theta-\beta)^*\xi]^0\mu(d\beta) \\
&\qquad +(e^{D^*(t-s)}\psi)(\theta),\ \theta\in[-\delta,0].
\end{aligned}$$

Notice that

$$\begin{aligned}
&\int_t^T \mathbf{T}(r,t)^*\mathbf{Q}(r)\mathbf{T}_\Theta(r,t)zdr \\
&= \int_t^T \mathbf{T}(r,t)^* \begin{bmatrix} \tilde{Q}(r)[\mathbf{T}_\Theta(r,t)z]^0 + \tilde{S}_1(r)^*[\mathbf{T}_\Theta(r,t)z]^1 \\ \tilde{S}_1(r)[\mathbf{T}_\Theta(r,t)z]^0 + \tilde{R}_{11}(r)[\mathbf{T}_\Theta(r,t)z]^1 \end{bmatrix} dr = \begin{pmatrix} P_0(t)z \\ P_1(t)z \end{pmatrix},
\end{aligned}$$

where

$$\begin{aligned}
P_0(t)z &:= \int_t^T \Phi(r,t)^*\Big[\tilde{Q}(r)[\mathbf{T}_\Theta(r,t)z]^0 + \tilde{S}_1(r)^*[\mathbf{T}_\Theta(r,t)z]^1\Big]dr, \\
(P_1(t)z)(\theta) &:= \int_t^T \bigg\{\Big(e^{D^*(r-t)}\Big[\tilde{S}_1(r)[\mathbf{T}_\Theta(r,t)z]^0 + \tilde{R}_{11}(r)[\mathbf{T}_\Theta(r,t)z]^1\Big]\Big)(\theta) \\
&\quad + \int_{[t+\theta-r,\theta]} B(t+\theta-\beta,\beta)^\top\Big\{\Phi(r,t+\theta \\
&\quad -\beta)^*\Big[\tilde{Q}(r)[\mathbf{T}_\Theta(r,t)z]^0 \\
&\quad +\tilde{S}_1(r)^*[\mathbf{T}_\Theta(r,t)z]^1\Big]\Big\}^0\mu(d\beta)\bigg\}dr.
\end{aligned}$$

By the continuity of $B(\cdot,\cdot)$ and the strong continuity of $\Phi(\cdot,\cdot)^*$, $P_1(t)z$ is meaningful. Next we aim to prove that $\mathcal{D}^*P_1(t)h$ is well-defined. Recalling Lemma 3.2.5, $\mathcal{D}^*$ can be defined in the generalized sense. Noting $P_1(t)z \in L^2$, $B(\cdot,\cdot)$, $S_{10}(\cdot,\cdot)$, $S_{11}(\cdot,\cdot,\cdot)$, $R_{11}(\cdot,\cdot,\cdot)$ are continuous and $\mu((\theta,0]) \to 0$, as $\theta \to 0$, $P_1(t)z$ is continuous at $\theta = 0$, hence $\mathbf{B}^* P(t)$ is well-defined. From the proof of Lemma 3.2.7, $\mathbf{T}^\lambda_\Theta(t, s)$ is uniformly bounded in λ, t, s. By Lemma 3.2.5, we have

$$
\begin{aligned}
&\Big|(\mathcal{D}^\lambda)^*\Big[\int_t^T \mathbf{T}(r,t)^*\mathbf{Q}(r)\mathbf{T}^\lambda_\Theta(r,t)zdr\Big]^1 \\
&\quad - \mathcal{D}^*\Big[\int_t^T \mathbf{T}(r,t)^*\mathbf{Q}(r)\mathbf{T}_\Theta(r,t)zdr\Big]^1\Big| \\
&= \Big|(\mathcal{D}^\lambda)^*\Big[\int_t^T \mathbf{T}(r,t)^*\mathbf{Q}(r)(\mathbf{T}^\lambda_\Theta(r,t) - \mathbf{T}_\Theta(r,t))zdr\Big]^1 \\
&\quad - \big[(\mathcal{D}^\lambda)^* - \mathcal{D}^*\big]\Big[\int_t^T \mathbf{T}(r,t)^*\mathbf{Q}(r)\mathbf{T}_\Theta(r,t)zdr\Big]^1\Big| \to 0, \qquad \text{as } \lambda \to 0,
\end{aligned}
$$

hence $\mathbf{B}^{\lambda *} P^\lambda(t)z \to \mathbf{B}^* P(t)z$ as $\lambda \to \infty$. Finally by the uniform boundedness of $\mathbf{T}_\Theta, \Phi, \tilde{Q}, \tilde{S}_1, \tilde{R}_{11}, B$, we conclude that $\sup_{s\leqslant t\leqslant T} \|\mathbf{B}^* P(t)\|_{\mathscr{L}(Z,\mathbb{R}^m)} < \infty$. This completes the proof of Lemma 3.2.8. □

By (3.2.11), Lemmas 3.2.7 and 3.2.8, we derive

$$\Theta(t, s) = -\mathbf{R}(t)^{-1}\mathbf{B}^* P(t)\mathbf{T}_\Theta(t, s), \tag{3.2.29}$$

hence we obtain

$$\mathbf{T}_\Theta(t, s)\mathbf{Z}_0 = \mathbf{T}(t, s)\mathbf{Z}_0 - \int_s^t \mathbf{T}(t, r)\mathbf{B}\mathbf{R}(r)^{-1}\mathbf{B}^* P(r)\mathbf{T}_\Theta(r, s)\mathbf{Z}_0 dr.$$

Since $R_{00}(\cdot)$ is continuous, $\mathbf{R}^{-1}\mathbf{B}^* P$ is strongly continuous. By Lemma 2.2.5, we conclude that the above integral equation admits the unique solution $\mathbf{T}_\Theta(\cdot, \cdot)$, which is a mild evolution operator.

Now we can give the closed-loop representation of open-loop optimal control for Problem (P_0).

Theorem 3.2.9 *Let* (H3.2) *hold. Denote* $\mathbf{B}^* := (0, \mathcal{D}^*)$ *(where* $\mathcal{D}^*$ *is understood in the generalized sense, see Lemma 3.2.5). Assume* μ *is absolutely continuous in Lebesgue measure, then* $P^\lambda(t)$ *converges strongly to a strongly continuous nonnegative operator* $P(t) \in \mathscr{L}(Z)$ *and* $\mathbf{B}^* P(t) \in \mathscr{L}(Z, \mathbb{R}^m)$ *is well-defined. Moreover,* $\mathbf{B}^{\lambda *} P^\lambda(t)$ *converges strongly to* $\mathbf{B}^* P(t)$*, and there exists a constant* $K > 0$

such that $\|\mathbf{B}^*P(t)\|_{\mathscr{L}(Z,\mathbb{R}^m)} \leqslant K$ *for all* $t \in [s, T]$. *Furthermore, for any* $z \in Z$, $P(\cdot)$ *satisfies the following integral Riccati equations:*

$$
\begin{aligned}
(a)\quad & P(t)z = \mathbf{T}(T,t)^*\mathbf{G}\mathbf{T}(T,t)z + \int_t^T \mathbf{T}(r,t)^*\Big\{\mathbf{Q}(r) - \big[\mathbf{B}^*P(r)\big]^* \\
& \qquad \times \mathbf{R}(r)^{-1}\big[\mathbf{B}^*P(r)\big]\Big\}\mathbf{T}(r,t)z\,dr, \quad t \in [s,T], \\
(b)\quad & P(t)z = \mathbf{T}(T,t)^*\mathbf{G}\mathbf{T}_\Theta(T,t)z + \int_t^T \mathbf{T}(r,t)^*\mathbf{Q}(r)\mathbf{T}_\Theta(r,t)z\,dr, \\
& \qquad t \in [s,T], \\
(c)\quad & P(t)z = \mathbf{T}_\Theta(T,t)^*\mathbf{G}\mathbf{T}_\Theta(T,t)z + \int_t^T \mathbf{T}_\Theta(r,t)^*\Big\{\mathbf{Q}(r) + \big[\mathbf{B}^*P(r)\big]^* \\
& \qquad \times \mathbf{R}(r)^{-1}\big[\mathbf{B}^*P(r)\big]\Big\}\mathbf{T}_\Theta(r,t)z\,dr, \, t \in [s,T],
\end{aligned}
\tag{3.2.30}
$$

where $\mathbf{T}_\Theta(\cdot\,,\cdot)$ *is the solution to the following integral equation:*

$$
\mathbf{T}_\Theta(t,s)z = \mathbf{T}(t,s)z - \int_s^t \mathbf{T}(t,r)\mathbf{B}\mathbf{R}(r)^{-1}\mathbf{B}^*P(r)\mathbf{T}_\Theta(r,s)z\,dr.
$$

Then the optimal control of Problem (P_0) *is*

$$
\bar{u}(t) = -\mathbf{R}(t)^{-1}\mathbf{B}^*P(t)\bar{\mathbf{Z}}(t), \quad \text{a.e. } t \in [s,T], \tag{3.2.31}
$$

and the value function is

$$
V(s,\mathbf{Z}_0) = \big\langle P(s)\mathbf{Z}_0, \mathbf{Z}_0\big\rangle_Z. \tag{3.2.32}
$$

Proof Recall the equations in (3.2.3) are equivalent to each other, as (3.2.3)(a), (3.2.3)(b), (3.2.3)(c) converge weakly to (3.2.30)(a), (3.2.30)(c), (3.2.30)(b), respectively, the equations in (3.2.30) are also equivalent to each other. By (3.2.29) and Lemma 3.2.6, we conclude that (3.2.31) is the optimal control of Problem (P_0). The strong continuity of $P(t)$ follows from the strong continuity of $P^\lambda(t)$ by Theorem 3.2.1, which completes the proof. □

Theorem 3.2.9 gives the closed-loop representation of open-loop optimal control for Problem (P_0) by Problem (EP_0). Next we try to give a more straightforward representation by removing those abstract operators.

In the rest of this section, consider the special state Eq. (3.1.13) with $b(\cdot) = 0$ instead of (2.1.1). Introduce the closed linear operator $\mathbf{A}(t)$ associated with $\mathbf{T}(t, s)$ as follows:

$$\mathbf{A}(t) : \mathscr{D}(\mathbf{A}(t)) \longrightarrow Z$$
$$\mathbf{A}(t) := \begin{bmatrix} \tilde{A}(t) & \widetilde{B}(t) \\ 0 & D \end{bmatrix},$$

where $\tilde{A}(t)$, $\widetilde{B}(t)$ and D are defined by (3.1.15), (3.1.16) and (2.1.19), respectively. Recall $\mathscr{W} = \mathscr{D}(\tilde{A}(t))$ and $W = \mathscr{D}(D)$, denote $\mathscr{W}W := \mathscr{D}(\mathbf{A}(t)) = \mathscr{W} \times W$. Using the properties of $\tilde{A}(t)$ and D (refer to [2, 3, 7]), we have

$$\begin{aligned} &\frac{\partial}{\partial t}\mathbf{T}(t, s)\mathbf{Z} = \mathbf{A}(t)\mathbf{T}(t, s)\mathbf{Z}, \quad \text{a.e. } t \in [s, T], \ \forall\, \mathbf{Z} \in \mathscr{W}W, \\ &\frac{\partial}{\partial s}\mathbf{T}(t, s)\mathbf{Z} = -\mathbf{T}(t, s)\mathbf{A}(s)\mathbf{Z}, \quad \text{a.e. } s \in [0, t], \ \forall\, \mathbf{Z} \in \mathscr{W}W. \end{aligned} \tag{3.2.33}$$

In fact, for any $\mathbf{Z} = \begin{pmatrix} \xi \\ \psi \end{pmatrix} \in \mathscr{W}W$, notice that

$$\begin{aligned} \frac{\partial}{\partial s}\mathbf{T}(t, s)\mathbf{Z} &= \frac{\partial}{\partial s}\begin{bmatrix} \Phi(t, s)\xi + \int_s^t \Phi(t, r)\widetilde{B}(r)e^{D(r-s)}\psi dr \\ \varepsilon^{D(t-s)}\psi \end{bmatrix} \\ &= \begin{bmatrix} -\Phi(t, s)\tilde{A}(s)\xi - \Phi(t, s)\widetilde{B}(s)\psi - \int_s^t \Phi(t, r)\widetilde{B}(r)e^{D(r-s)}D\psi dr \\ \varepsilon^{D(t-s)}D\psi \end{bmatrix} \\ &= -\mathbf{T}(t, s)\mathbf{A}(s)\mathbf{Z}, \end{aligned}$$

then the second equality of (3.2.33) holds, and the first equality can be verified similarly.

Notice that (3.2.30) also satisfies the following differential Riccati equation:

$$\begin{cases} \frac{d}{dt}\langle P(t)z, z'\rangle_Z + \langle P(t)z, \mathbf{A}(t)z'\rangle_Z + \langle \mathbf{A}(t)z, P(t)z'\rangle_Z \\ +\langle \mathbf{Q}(t)z, z'\rangle_Z - \langle [\mathbf{B}^* P(t)]^* \mathbf{R}(t)^{-1}[\mathbf{B}^* P(t)]z, z'\rangle_Z = 0, \\ \qquad\qquad \text{a.e. } t \in [s, T], \quad \forall\, z, z' \in \mathscr{W}W, \\ P(T) = \mathbf{G}. \end{cases} \tag{3.2.34}$$

To obtain a straightforward representation for the closed-loop representation of open-loop optimal control for Problem (P_0), next we give a heuristic derivation. Let us decompose $P(t)$ as follows:

$$P(t)z = \begin{bmatrix} P_{00}(t) & P_{01}(t) \\ P_{10}(t) & P_{11}(t) \end{bmatrix} \begin{bmatrix} \xi \\ \psi \end{bmatrix}, \quad \forall\, z = \begin{bmatrix} \xi \\ \psi \end{bmatrix} \in Z,$$

where $P_{00}(t) \in \mathscr{L}(M^2)$, $P_{01}(t) \in \mathscr{L}(L^2, M^2)$, $P_{10}(t) \in \mathscr{L}(M^2, L^2)$, $P_{11}(t) \in \mathscr{L}(L^2)$ and $P_{01}(t)^* = P_{10}(t)$. Then from (3.2.34) we can derive (for simplicity we use $\langle \cdot, \cdot \rangle$ substituting $\langle \cdot, \cdot \rangle_{M^2}$ and $\langle \cdot, \cdot \rangle_{L^2}$ if no ambiguity exists)

$$\begin{cases} (a)\ \dfrac{d}{dt}\big\langle P_{00}(t)\xi, \xi' \big\rangle + \big\langle P_{00}(t)\xi, \tilde{A}(t)\xi' \big\rangle + \big\langle \tilde{A}(t)\xi, P_{00}(t)\xi' \big\rangle \\ \qquad + \big\langle \tilde{Q}(t)\xi, \xi' \big\rangle - \big\langle \xi', (\mathcal{D}^* P_{10}(t))^* R_{00}(t)^{-1} (\mathcal{D}^* P_{10}(t))\xi \big\rangle = 0, \\ \qquad\qquad\qquad\qquad \text{a.e. } t \in [s, T],\ \forall\, \xi, \xi' \in \mathscr{W}, \\ (b)\ \dfrac{d}{dt}\big\langle P_{10}(t)\xi, \psi' \big\rangle + \big\langle P_{00}(t)\xi, \widetilde{B}(t)\psi' \big\rangle + \big\langle D\psi', P_{10}(t)\xi \big\rangle \\ \qquad + \big\langle \tilde{A}(t)\xi, P_{01}(t)\psi' \big\rangle + \big\langle \tilde{S}_1(t)\xi, \psi' \big\rangle \\ \qquad - \big\langle \psi', (\mathcal{D}^* P_{11}(t))^* R_{00}(t)^{-1} (\mathcal{D}^* P_{10}(t))\xi \big\rangle = 0, \\ \qquad\qquad\qquad\qquad \text{a.e. } t \in [s, T],\ \forall\, \xi \in \mathscr{W},\ \psi' \in W, \\ (c)\ \dfrac{d}{dt}\big\langle P_{11}\psi, \psi' \big\rangle + \big\langle P_{01}(t)\psi, \widetilde{B}(t)\psi' \big\rangle + \big\langle D\psi', P_{11}(t)\psi \big\rangle \\ \qquad + \big\langle \widetilde{B}(t)\psi, P_{01}(t)\psi' \big\rangle + \big\langle D\psi, P_{11}(t)\psi' \big\rangle + \big\langle \tilde{R}_{11}(t)\psi, \psi' \big\rangle \\ \qquad - \big\langle \psi', (\mathcal{D}^* P_{11}(t))^* R_{00}(t)^{-1} (\mathcal{D}^* P_{11}(t))\psi \big\rangle = 0, \\ \qquad\qquad\qquad\qquad \text{a.e. } t \in [s, T],\ \forall\, \psi,\ \psi' \in W, \\ P_{00}(T) = \tilde{G},\ P_{10}(T) = 0,\ P_{11}(T) = 0, \quad P_{01}(t)^* = P_{10}(t). \end{cases} \tag{3.2.35}$$

Suppose $P_{00}(t)$ and $P_{11}(t)$ can be decomposed into the following form:

$$P_{00}(t)\xi = \begin{bmatrix} P_{00}^{(00)}(t) & P_{00}^{(01)}(t) \\ P_{00}^{(10)}(t) & P_{00}^{(11)}(t) \end{bmatrix} \begin{bmatrix} x \\ \varphi \end{bmatrix}, \quad \forall\, \xi = \begin{bmatrix} x \\ \varphi \end{bmatrix} \in M^2,$$

$$P_{10}(t)\xi = (P_{10}^{(00)}(t), P_{10}^{(01)}(t)) \begin{bmatrix} x \\ \varphi \end{bmatrix}, \quad \forall\, \xi = \begin{bmatrix} x \\ \varphi \end{bmatrix} \in M^2,$$

where $P^{(00)}_{00}(t) \in \mathbb{R}^{n\times n}$, $P^{(01)}_{00}(t) \in \mathscr{L}(L^2, \mathbb{R}^n)$, $P^{(10)}_{00}(t) \in \mathscr{L}(\mathbb{R}^n, L^2)$, $P^{(11)}_{00}(t) \in \mathscr{L}(L^2)$, $P^{(00)}_{10}(t) \in \mathscr{L}(\mathbb{R}^n, L^2)$, $P^{(01)}_{10}(t) \in \mathscr{L}(L^2, L^2)$. Furthermore assume that there exists $\mathbb{R}^{n\times n}$-valued functions $\mathcal{E}_1(\cdot,\cdot), \mathcal{E}_2(\cdot,\cdot,\cdot)$, and $\mathbb{R}^{m\times n}$-valued functions $\mathcal{E}_3(\cdot,\cdot), \mathcal{E}_4(\cdot,\cdot,\cdot)$ such that

$$
\begin{aligned}
&[P^{(10)}_{00}(t)x](\theta) = \mathcal{E}_1(t,\theta)x, \quad [P^{(00)}_{10}(t)x](\theta) = \mathcal{E}_3(t,\theta)x, \quad \forall\, x \in \mathbb{R}^n,\\
&[P^{(11)}_{00}(t)\varphi](\theta) = \int_{-\delta}^{0} \mathcal{E}_2(t,\theta,\alpha)\varphi(\alpha)d\alpha,\\
&[P^{(01)}_{10}(t)\varphi](\theta) = \int_{-\delta}^{0} \mathcal{E}_4(t,\theta,\alpha)\varphi(\alpha)d\alpha, \quad \forall\, \varphi \in L^2,
\end{aligned}
$$

and $\mathcal{E}_2(t,\theta,\alpha)^\top = \mathcal{E}_2(t,\alpha,\theta)$. For simplicity of writing, denote $\mathcal{E}_0(t) := P^{(00)}_{00}(t)$, then (3.2.35)(a) can be decomposed as the following *coupled matrix-valued Riccati equations* (CMREs, for short):

$$
\begin{cases}
\frac{d}{dt}\mathcal{E}_0(t) + A_0(t)^\top\mathcal{E}_0(t) + \mathcal{E}_0(t)A_0(t) + Q_{00}(t) + \mathcal{E}_1(t,0) + \mathcal{E}_1(t,0)^\top \\
-\big[B_0(t)^\top\mathcal{E}_0(t) + \mathcal{E}_3(t,0)\big]^\top R_{00}(t)^{-1}\big[B_0(t)^\top\mathcal{E}_0(t) + \mathcal{E}_3(t,0)\big] = 0, \text{ a.e. } t,\\
\Big(\frac{\partial}{\partial t} - \frac{\partial}{\partial\theta}\Big)\mathcal{E}_1(t,\theta) + \mathcal{E}_1(t,\theta)A_0(t) + A^0(t,\theta)^\top\mathcal{E}_0(t) + \mathcal{E}_2(t,\theta,0)\\
+Q_{10}(t,\theta) - \big[\mathcal{E}_1(t,\theta)B_0(t) + \mathcal{E}_4(t,0,\theta)^\top\big]R_{00}(t)^{-1}\big[B_0(t)^\top\mathcal{E}_0(t)\\
+\mathcal{E}_3(t,0)\big] + \sum_{i=1}^{N-1} A_i(t)^\top\mathcal{E}_0(t)\hat{\delta}(\theta-\theta_i) = 0, \quad \text{a.e. } t,\theta,\\
\Big(\frac{\partial}{\partial t} - \frac{\partial}{\partial\theta} - \frac{\partial}{\partial\alpha}\Big)\mathcal{E}_2(t,\theta,\alpha) + A^0(t,\theta)^\top\mathcal{E}_1(t,\alpha)^\top\\
+\mathcal{E}_1(t,\theta)A^0(t,\alpha) + Q_{11}(t,\alpha,\theta) - \big[B_0(t)^\top\mathcal{E}_1(t,\theta)^\top\\
+\mathcal{E}_4(t,0,\theta)\big]^\top R_{00}(t)^{-1}\big[B_0(t)^\top\mathcal{E}_1(t,\alpha)^\top + \mathcal{E}_4(t,0,\alpha)\big]\\
+\sum_{i=1}^{N-1}\Big[A_i(t)^\top\mathcal{E}_1(t,\alpha)^\top\hat{\delta}(\theta-\theta_i) + \mathcal{E}_1(t,\theta)A_i(t)\hat{\delta}(\alpha-\theta_i)\Big] = 0,\\
\qquad\qquad \text{a.e. } t,\theta,\alpha,\\
\mathcal{E}_0(T) = G_{00},\ \mathcal{E}_1(t,-\delta) = A_N(t)^\top\mathcal{E}_0(t),\ \mathcal{E}_1(T,\theta) = G_{10}(\theta),\\
\mathcal{E}_2(t,\alpha,-\delta) = \mathcal{E}_1(t,\alpha)A_N(t),\ \mathcal{E}_2(t,-\delta,\alpha) = A_N(t)^\top\mathcal{E}_1(t,\alpha)^\top,\\
\mathcal{E}_2(T,\theta,\alpha) = G_{11}(\alpha,\theta),
\end{cases}
\tag{3.2.36}
$$

(3.2.35)(b) can be decomposed as:

$$
\begin{cases}
\Big(\frac{\partial}{\partial t}-\frac{\partial}{\partial \theta}\Big)\mathcal{E}_3(t,\theta)+\Big[\sum_{i=1}^{N-1}B_i(t)^\top\hat{\delta}(\theta-\theta_i)+B^0(t,\theta)^\top\Big]\mathcal{E}_0(t)\\
\quad+\mathcal{E}_3(t,\theta)A_0(t)+\mathcal{E}_4(t,\theta,0)+S_{10}(t,\theta)-\big[\mathcal{E}_3(t,\theta)B_0(t)\\
\quad+\mathcal{E}_5(t,\theta,0)\big]R_{00}(t)^{-1}\big[B_0(t)^\top\mathcal{E}_0(t)+\mathcal{E}_3(t,0)\big]=0, \quad \text{a.e. } t,\theta,\\
\Big(\frac{\partial}{\partial t}-\frac{\partial}{\partial \theta}-\frac{\partial}{\partial \alpha}\Big)\mathcal{E}_4(t,\theta,\alpha)+\Big[\sum_{i=1}^{N-1}B_i(t)^\top\hat{\delta}(\theta-\theta_i)+B^0(t,\theta)^\top\Big]\\
\quad\times\mathcal{E}_1(t,\alpha)^\top+\mathcal{E}_3(t,\theta)\Big[\sum_{i=1}^{N-1}A_i(t)\hat{\delta}(\alpha-\theta_i)+A^0(t,\alpha)\Big]\\
\quad+S_{11}(t,\alpha,\theta)-\big[\mathcal{E}_3(t,\theta)B_0(t)+\mathcal{E}_5(t,\theta,0)\big]\\
\quad\times R_{00}(t)^{-1}\big[B_0(t)^\top\mathcal{E}_1(t,\alpha)^\top+\mathcal{E}_4(t,0,\alpha)\big]=0, \quad \text{a.e. } t,\theta,\alpha,\\
\mathcal{E}_3(t,-\delta)=B_N(t)^\top\mathcal{E}_0(t), \quad \mathcal{E}_3(T,\theta)=0, \quad \mathcal{E}_4(T,\theta,\alpha)=0,\\
\mathcal{E}_4(t,-\delta,\alpha)=B_N(t)^\top\mathcal{E}_1(t,\alpha)^\top, \ \mathcal{E}_4(t,\theta,-\delta)=\mathcal{E}_3(t,\theta)A_N(t),
\end{cases}
\tag{3.2.37}
$$

and (3.2.35)(c) can be decomposed as:

$$
\begin{cases}
\Big(\frac{\partial}{\partial t}-\frac{\partial}{\partial \theta}-\frac{\partial}{\partial \alpha}\Big)\mathcal{E}_5(t,\theta,\alpha)+\Big[\sum_{i=1}^{N-1}B_i(t)^\top\hat{\delta}(\theta-\theta_i)+B^0(t,\theta)^\top\Big]\\
\quad\times\mathcal{E}_3(t,\alpha)^\top+\mathcal{E}_3(t,\theta)\Big[\sum_{i=1}^{N-1}B_i(t)\hat{\delta}(\alpha-\theta_i)+B^0(t,\alpha)\Big]\\
\quad+R_{11}(t,\alpha,\theta)-\big[\mathcal{E}_3(t,\theta)B_0(t)+\mathcal{E}_5(t,\theta,0)\big]R_{00}(t)^{-1}\\
\quad\times\big[B_0(t)^\top\mathcal{E}_3(t,\alpha)^\top+\mathcal{E}_5(t,0,\alpha)\big]=0, \quad \text{a.e. } t,\theta,\alpha,\\
\mathcal{E}_5(t,\theta,-\delta)=\mathcal{E}_3(t,\theta)B_N(t), \ \mathcal{E}_5(t,-\delta,\theta)=B_N(t)^\top\mathcal{E}_3(t,\theta)^\top,\\
\mathcal{E}_5(T,\theta,\alpha)=0.
\end{cases}
\tag{3.2.38}
$$

Then the optimal control (3.2.31) can be written as

$$
\begin{aligned}
\bar{u}(t)=&-R_{00}(t)^{-1}\Big\{\big[B_0(t)^\top\mathcal{E}_0(t)+\mathcal{E}_3(t,0)\big]\bar{X}(t)+\int_{-\delta}^{0}\big[B_0(t)^\top\mathcal{E}_1(t,\theta)^\top\\
&+\mathcal{E}_4(t,0,\theta)\big]\bar{X}(t+\theta)d\theta+\int_{-\delta}^{0}\big[B_0(t)^\top\mathcal{E}_3(t,\theta)^\top\\
&+\mathcal{E}_5(t,0,\theta)\big]\bar{u}(t+\theta)d\theta\Big\}, \quad \text{a.e. } t\in[s,T],
\end{aligned}
\tag{3.2.39}
$$

and the value function (3.2.32) can be written as

$$
\begin{aligned}
J(s,\mathbf{Z}_0;\bar{u}(\cdot)) &= \big\langle \mathcal{E}_0(s)x, x\big\rangle + 2\int_{-\delta}^{0}\big\langle \mathcal{E}_1(s,\theta)x, \varphi(\theta)\big\rangle d\theta \\
&+ \int_{-\delta}^{0}\int_{-\delta}^{0}\big\langle \mathcal{E}_2(s,\theta,\alpha)\varphi(\alpha), \varphi(\theta)\big\rangle d\alpha d\theta + 2\int_{-\delta}^{0}\big\langle \mathcal{E}_3(s,\theta)x, \psi(\theta)\big\rangle d\theta \\
&+2\int_{-\delta}^{0}\int_{-\delta}^{0}\big\langle \mathcal{E}_4(s,\theta,\alpha)\varphi(\alpha), \psi(\theta)\big\rangle d\alpha d\theta \\
&+ \int_{-\delta}^{0}\int_{-\delta}^{0}\big\langle \mathcal{E}_5(s,\theta,\alpha)\psi(\alpha), \psi(\theta)\big\rangle d\alpha d\theta.
\end{aligned}
\tag{3.2.40}
$$

Now we can give the following theorem to explicitly express the optimal control for Problem (P_0).

Theorem 3.2.10 *Let* (H3.1) *hold. Assume* $R_{00} > 0$, S_{00}, S_{01}, $R_{10} = 0$. *Let* $\mathcal{E}_0(t)$, $\mathcal{E}_1(t,\theta)$, $\mathcal{E}_2(t,\theta,\alpha)$, $\mathcal{E}_3(t,\theta)$, $\mathcal{E}_4(t,\theta,\alpha)$, $\mathcal{E}_5(t,\theta,\alpha)$, $t\in[s,T]$, $\theta,\alpha\in[-\delta,0]$, *satisfy almost everywhere the CMREs (3.2.36)–(3.2.38), and* $\mathcal{E}_0(t) = \mathcal{E}_0(t)^\top$, $\mathcal{E}_2(t,\theta,\alpha) = \mathcal{E}_2(t,\alpha,\theta)^\top$, $\mathcal{E}_5(t,\theta,\alpha) = \mathcal{E}_5(t,\alpha,\theta)^\top$, *then the optimal control for* Problem (P_0), *with the state Eq.* (3.1.13) *instead of* (2.1.1), *is (3.2.39), and the value function is (3.2.40).*

Proof For any $u(\cdot)\in L^2(s,T;\mathbb{R}^m)$, let $X(\cdot)$ satisfy the following equation:

$$
\begin{cases}
\dot{X}(t) = \displaystyle\sum_{i=0}^{N} A_i(t)X(t+\theta_i) + \int_{-\delta}^{0} A^0(t,\theta)X(t+\theta)d\theta \\
\qquad + \displaystyle\sum_{i=0}^{N} B_i(t)u(t+\theta_i) + \int_{-\delta}^{0} B^0(t,\theta)u(t+\theta)d\theta, \qquad \text{a.e. } t\in[s,T], \\
X(s) = x, \quad X(t) = \varphi(t-s), \quad t\in[s-\delta,s), \\
u(t) = \psi(t-s), \quad t\in[s-\delta,s).
\end{cases}
$$

Define

$$
\begin{aligned}
\Gamma(t) &:= \big\langle \mathcal{E}_0(t)X(t), X(t)\big\rangle + 2\int_{-\delta}^{0}\big\langle \mathcal{E}_1(t,\theta)X(t), X(t+\theta)\big\rangle d\theta \\
&+ \int_{-\delta}^{0}\int_{-\delta}^{0}\big\langle \mathcal{E}_2(t,\theta,\alpha)X(t+\alpha), X(t+\theta)\big\rangle d\alpha d\theta \\
&+2\int_{-\delta}^{0}\big\langle \mathcal{E}_3(t,\theta)X(t), u(t+\theta)\big\rangle d\theta \\
&+2\int_{-\delta}^{0}\int_{-\delta}^{0}\big\langle \mathcal{E}_4(t,\theta,\alpha)X(t+\alpha), u(t+\theta)\big\rangle d\alpha d\theta \\
&+ \int_{-\delta}^{0}\int_{-\delta}^{0}\big\langle \mathcal{E}_5(t,\theta,\alpha)u(t+\alpha), u(t+\theta)\big\rangle d\alpha d\theta,
\end{aligned}
$$

then we have (with some computations)

$$
\begin{aligned}
&\frac{d}{dt}\Gamma(t) + \big\langle Q_{00}(t)X(t), X(t)\big\rangle + 2\int_{-\delta}^{0}\big\langle Q_{10}(t,\theta)^\top X(t+\theta), X(t)\big\rangle d\theta \\
&+\int_{-\delta}^{0}\int_{-\delta}^{0}\big\langle Q_{11}(t,\theta,\theta')X(t+\theta), X(t+\theta')\big\rangle d\theta' d\theta \\
&+2\int_{-\delta}^{0}\big\langle S_{10}(t,\theta)^\top u(t+\theta), X(t)\big\rangle d\theta \\
&+2\int_{-\delta}^{0}\int_{-\delta}^{0}\big\langle S_{11}(t,\theta,\theta')X(t+\theta), u(t+\theta')\big\rangle d\theta' d\theta + \big\langle R_{00}(t)u(t), u(t)\big\rangle \\
&+\int_{-\delta}^{0}\int_{-\delta}^{0}\big\langle R_{11}(t,\theta,\theta')u(t+\theta), u(t+\theta')\big\rangle d\theta' d\theta \\
=&\Big\langle R_{00}(t)\Big\{u(t) + R_{00}(t)^{-1}\Big[\big[B_0(t)^\top \mathcal{E}_0(t) + \mathcal{E}_3(t,0)\big]X(t) \\
&\quad+\int_{-\delta}^{0}\big[B_0(t)^\top \mathcal{E}_1(t,\theta)^\top + \mathcal{E}_4(t,0,\theta)\big]X(t+\theta)d\theta \\
&\quad+\int_{-\delta}^{0}\big[B_0(t)^\top \mathcal{E}_3(t,\theta)^\top + \mathcal{E}_5(t,0,\theta)\big]u(t+\theta)d\theta\Big]\Big\}, \\
&\quad u(t) + R_{00}(t)^{-1}\Big[\big[B_0(t)^\top \mathcal{E}_0(t) + \mathcal{E}_3(t,0)\big]X(t) \\
&\quad+\int_{-\delta}^{0}\big[B_0(t)^\top \mathcal{E}_1(t,\theta)^\top + \mathcal{E}_4(t,0,\theta)\big]X(t+\theta)d\theta \\
&\quad+\int_{-\delta}^{0}\big[B_0(t)^\top \mathcal{E}_3(t,\theta)^\top + \mathcal{E}_5(t,0,\theta)\big]u(t+\theta)d\theta\Big]\Big\rangle, \quad \text{a.e. } t \in [s,T].
\end{aligned}
\tag{3.2.41}
$$

Substituting (3.2.39) into (3.2.41) and then integrating from s to T, we deduce

$$
\begin{aligned}
\Gamma(s) =& \big\langle G_{00}\bar{X}(T), \bar{X}(T)\big\rangle + 2\int_{-\delta}^{0}\big\langle G_{10}(\theta)^\top \bar{X}(T+\theta), \bar{X}(T)\big\rangle d\theta \\
&+\int_{-\delta}^{0}\int_{-\delta}^{0}\big\langle G_{11}(\theta,\theta')\bar{X}(T+\theta), \bar{X}(T+\theta')\big\rangle d\theta' d\theta \\
&+\int_{s}^{T}\Big\{\big\langle Q_{00}(t)\bar{X}(t), \bar{X}(t)\big\rangle + 2\int_{-\delta}^{0}\big\langle Q_{10}(t,\theta)^\top \bar{X}(t+\theta), \bar{X}(t)\big\rangle d\theta \\
&+\int_{-\delta}^{0}\int_{-\delta}^{0}\big\langle Q_{11}(t,\theta,\theta')\bar{X}(t+\theta), \bar{X}(t+\theta')\big\rangle d\theta' d\theta \\
&+2\int_{-\delta}^{0}\big\langle S_{10}(t,\theta)^\top \bar{u}(t+\theta), \bar{X}(t)\big\rangle d\theta + \big\langle R_{00}(t)\bar{u}(t), \bar{u}(t)\big\rangle
\end{aligned}
$$

$$
\begin{aligned}
&+2\int_{-\delta}^{0}\int_{-\delta}^{0}\big\langle S_{11}(t,\theta,\theta')\bar{X}(t+\theta),\bar{u}(t+\theta')\big\rangle d\theta' d\theta\\
&+\int_{-\delta}^{0}\int_{-\delta}^{0}\big\langle R_{11}(t,\theta,\theta')\bar{u}(t+\theta),\bar{u}(t+\theta')\big\rangle d\theta' d\theta\Big\}dt\\
&= J_0(s,x,\varphi(\cdot),\psi(\cdot);\bar{u}(\cdot)),
\end{aligned}
\tag{3.2.42}
$$

and noting $R_{00} > 0$, we have

$$
\begin{aligned}
\Gamma(s) \leqslant{}& \big\langle G_{00}X(T),X(T)\big\rangle + 2\int_{-\delta}^{0}\big\langle G_{10}(\theta)^\top X(T+\theta),X(T)\big\rangle d\theta\\
&+\int_{-\delta}^{0}\int_{-\delta}^{0}\big\langle G_{11}(\theta,\theta')X(T+\theta),X(T+\theta')\big\rangle d\theta' d\theta\\
&+\int_{s}^{T}\Big\{\big\langle Q_{00}(t)X(t),X(t)\big\rangle + 2\int_{-\delta}^{0}\big\langle Q_{10}(t,\theta)^\top X(t+\theta),X(t)\big\rangle d\theta\\
&\quad+\int_{-\delta}^{0}\int_{-\delta}^{0}\big\langle Q_{11}(t,\theta,\theta')X(t+\theta),X(t+\theta')\big\rangle d\theta' d\theta\\
&\quad+2\int_{-\delta}^{0}\big\langle S_{10}(t,\theta)^\top u(t+\theta),X(t)\big\rangle d\theta\\
&\quad+2\int_{-\delta}^{0}\int_{-\delta}^{0}\big\langle S_{11}(t,\theta,\theta')X(t+\theta),u(t+\theta')\big\rangle d\theta' d\theta\\
&\quad+\big\langle R_{00}(t)u(t),u(t)\big\rangle\\
&\quad+\int_{-\delta}^{0}\int_{-\delta}^{0}\big\langle R_{11}(t,\theta,\theta')u(t+\theta),u(t+\theta')\big\rangle d\theta' d\theta\Big\}dt\\
={}& J_0(s,x,\varphi(\cdot),\psi(\cdot);u(\cdot)), \quad \forall\, u(\cdot)\in L^2(s,T;\mathbb{R}^m),
\end{aligned}
$$

which and (3.2.42) imply that $\bar{u}(\cdot)$ is an optimal control for Problem (P_0) and the value function can be expressed as (3.2.40). This completes the proof of Theorem 3.2.10. □

Remark 3.2.11 Assume that $A_i, B_i = 0, i = 1, \cdots, N-1$, and $G_{00}, G_{10}, G_{11} = 0$. Then, the CMREs (3.2.36)–(3.2.38) admit unique solutions. The detailed proof can be referred to Alekal et al. [1].

Remark 3.2.12 Consider Problem (P_0) without the state delays, then when $A_0, B_i, B^0, Q_{00}, R_{00}$ are independent of time t and A_i, A^0, b, Q_{10}, Q_{11}, S_{00}, S_{01}, S_{10}, S_{11}, R_{10}, R_{11}, q_0, q_1, ρ_0, ρ_1, G_{10}, G_{11}, g_0, g_1 $= 0$, $i = 1, \cdots, N$, Theorem 3.2.9 reduces to Theorem 2.1 in Ichikawa [4].

Remark 3.2.13

(1) Consider Problem (P_0) without the control delays, $A^0, B_i, B^0, b, Q_{10}, Q_{11}, S_{00}, S_{01}, S_{10}, S_{11}, R_{10}, R_{11}, q_0, q_1, \rho_0, \rho_1, G_{00}, G_{10}, G_{11}, g_0, g_1 = 0, i = 1, \cdots, N$, $A_i = 0, i = 1, \cdots, N-1$, in this case $\mathcal{E}_3, \mathcal{E}_4, \mathcal{E}_5 = 0$, thus Theorem 3.2.10 reduces to Theorem 5 in Alekal et al. [1].

(2) Consider Problem (P_0) without the state delays, A_i, A^0, b, Q_{10}, Q_{11}, S_{00}, S_{01}, S_{10}, S_{11}, R_{10}, R_{11}, q_0, q_1, ρ_0, ρ_1, G_{10}, G_{11}, g_0, $g_1 = 0$, $i = 1, \cdots, N$, and the coefficients are time-invariant, then the CMREs (3.2.36)–(3.2.38) reduce to (2.29) in Ichikawa [4].
(3) Consider Problem (P_0) without the state delays and the control delays, then the CMREs (3.2.36)–(3.2.38) reduce to the classical Riccati equation in Yong-Zhou [13].

In the end of this section, we give the closed-loop representation of open-loop optimal control for Problem (Ex). Assume that no external target component enters the cycle absorber, and the desired conçentration of the target component in the outlet gas is 0. In this case $f \equiv 0$ and $a \equiv 0$.

Introduce the following CMREs:

$$
\begin{cases}
\frac{d}{dt}\mathcal{E}_0(t) - 2k_1\mathcal{E}_0(t) + q + 2\mathcal{E}_1(t,0) - r^{-1}\big[k_3\mathcal{E}_0(t) + \mathcal{E}_3(t,0)\big]^2 = 0, \\
\qquad\qquad\qquad\qquad\qquad\qquad \text{a.e. } t, \\
\Big(\frac{\partial}{\partial t} - \frac{\partial}{\partial \theta}\Big)\mathcal{E}_1(t,\theta) - k_1\mathcal{E}_1(t,\theta) + \mathcal{E}_2(t,\theta,0) \\
-\big[k_3\mathcal{E}_1(t,\theta) + \mathcal{E}_4(t,0,\theta)\big]r^{-1}\big[k_3\mathcal{E}_0(t) + \mathcal{E}_3(t,0)\big] = 0, \quad \text{a.e. } t,\theta, \\
\Big(\frac{\partial}{\partial t} - \frac{\partial}{\partial \theta} - \frac{\partial}{\partial \alpha}\Big)\mathcal{E}_2(t,\theta,\alpha) - r^{-1}\big[k_3\mathcal{E}_1(t,\theta) + \mathcal{E}_4(t,0,\theta)\big] \\
\times\big[k_3\mathcal{E}_1(t,\alpha) + \mathcal{E}_4(t,0,\alpha)\big] = 0, \quad \text{a.e. } t,\theta,\alpha, \\
\mathcal{E}_0(T) = 0, \ \mathcal{E}_1(t,-\delta) = k_2\mathcal{E}_0(t), \ \mathcal{E}_1(T,\theta) = 0, \\
\mathcal{E}_2(t,\alpha,-\delta) = k_2\mathcal{E}_1(t,\alpha), \ \mathcal{E}_2(t,-\delta,\alpha) = k_2\mathcal{E}_1(t,\alpha), \\
\mathcal{E}_2(T,\theta,\alpha) = 0,
\end{cases}
\tag{3.2.43}
$$

$$
\begin{cases}
\Big(\frac{\partial}{\partial t} - \frac{\partial}{\partial \theta}\Big)\mathcal{E}_3(t,\theta) - k_1\mathcal{E}_3(t,\theta) + \mathcal{E}_4(t,\theta,0) \\
-r^{-1}\big[k_3\mathcal{E}_3(t,\theta) + \mathcal{E}_5(t,\theta,0)\big]\big[k_3\mathcal{E}_0(t) + \mathcal{E}_3(t,0)\big] = 0, \quad \text{a.e. } t,\theta, \\
\Big(\frac{\partial}{\partial t} - \frac{\partial}{\partial \theta} - \frac{\partial}{\partial \alpha}\Big)\mathcal{E}_4(t,\theta,\alpha) - r^{-1}\big[k_3\mathcal{E}_3(t,\theta) + \mathcal{E}_5(t,\theta,0)\big] \\
\times\big[k_3\mathcal{E}_1(t,\alpha) + \mathcal{E}_4(t,0,\alpha)\big] = 0, \quad \text{a.e. } t,\theta,\alpha, \\
\mathcal{E}_3(t,-\delta) = k_4\mathcal{E}_0(t), \quad \mathcal{E}_3(T,\theta) = 0, \quad \mathcal{E}_4(T,\theta,\alpha) = 0, \\
\mathcal{E}_4(t,-\delta,\alpha) = k_4\mathcal{E}_1(t,\alpha), \ \mathcal{E}_4(t,\theta,-\delta) = k_2\mathcal{E}_3(t,\theta),
\end{cases}
\tag{3.2.44}
$$

and

$$\begin{cases} \Big(\dfrac{\partial}{\partial t}-\dfrac{\partial}{\partial \theta}-\dfrac{\partial}{\partial \alpha}\Big)\mathcal{E}_5(t,\theta,\alpha)-r^{-1}\big[k_3\mathcal{E}_3(t,\theta)+\mathcal{E}_5(t,\theta,0)\big] \\ \quad\times\big[k_3\mathcal{E}_3(t,\alpha)+\mathcal{E}_5(t,0,\alpha)\big]=0, \quad \text{a.e. } t,\theta,\alpha, \\ \mathcal{E}_5(t,\theta,-\delta)=k_4\mathcal{E}_3(t,\theta),\ \mathcal{E}_5(t,-\delta,\theta)=k_4\mathcal{E}_3(t,\theta), \\ \mathcal{E}_5(T,\theta,\alpha)=0. \end{cases} \tag{3.2.45}$$

Corollary 3.2.14 *Let $f\equiv 0$, $a\equiv 0$ and $r>0$. Let $\mathcal{E}_0(t)$, $\mathcal{E}_1(t,\theta)$, $\mathcal{E}_2(t,\theta,\alpha)$, $\mathcal{E}_3(t,\theta)$, $\mathcal{E}_4(t,\theta,\alpha)$, $\mathcal{E}_5(t,\theta,\alpha)$, $t\in[0,T]$, $\theta,\alpha\in[-\delta,0]$, satisfy almost everywhere the CMREs (3.2.43)–(3.2.45), then the optimal control for* Problem (Ex) *is*

$$\begin{aligned} \bar u(t)=-r^{-1}\Big\{&\big[k_3\mathcal{E}_0(t)+\mathcal{E}_3(t,0)\big]\bar X(t)+\int_{-\delta}^{0}\big[k_3\mathcal{E}_1(t,\theta) \\ &+\mathcal{E}_4(t,0,\theta)\big]\bar X(t+\theta)d\theta+\int_{-\delta}^{0}\big[k_3\mathcal{E}_3(t,\theta) \\ &+\mathcal{E}_5(t,0,\theta)\big]\bar u(t+\theta)d\theta\Big\}, \quad \text{a.e. } t\in[s,T], \end{aligned}$$

and the value function is (3.2.40).

3.3 Closed-Loop Solvability

In this section, we study the optimal control problem which involves only the state delays not the control delays. For its closed-loop solvability, we derive the sufficient and necessary conditions (see Theorem 3.3.9) by virtue of Problem (EP-NCD). Then Theorem 3.3.11 gives a more straight expression when Radon measure is used to characterize the discrete delays and the distributed delays.

First we reformulate the optimal control problem as follows. In this case the state Eq. (2.1.1) is simplified as the following ODDE:

$$\begin{cases} \dot X(t)=\displaystyle\int_{[-\delta,0]}A(t,\theta)X(t+\theta)\mu(d\theta)+B_0(t)u(t)+b(t), \ \text{a.e. } t\in[s,T], \\ X(s)=x, \quad X(t)=\varphi(t-s), \quad t\in[s-\delta,s). \end{cases} \tag{3.3.1}$$

The cost functional (2.2.1) becomes:

$$\begin{aligned} &J(s,x,\varphi(\cdot);u(\cdot)) \\ &= \int_s^T \Big[\big\langle Q_{00}(t)X(t), X(t)\big\rangle + 2\int_{-\delta}^0 \big\langle Q_{10}(t,\theta)^\top X(t+\theta), X(t)\big\rangle d\theta \\ &+ \int_{[-\delta,0]^2} \big\langle Q_{11}(t,\theta,\theta')X(t+\theta), X(t+\theta')\big\rangle d\theta' d\theta + \big\langle R_{00}(t)u(t), u(t)\big\rangle \\ &+2\big\langle S_{00}(t)X(t), u(t)\big\rangle + 2\int_{-\delta}^0 \big\langle S_{01}(t,\theta)X(t+\theta), u(t)\big\rangle d\theta \\ &+2\big\langle q_0(t), X(t)\big\rangle + 2\int_{-\delta}^0 \big\langle q_1(t,\theta), X(t+\theta)\big\rangle d\theta + 2\big\langle \rho_0(t), u(t)\big\rangle \Big] dt \\ &+\big\langle G_{00}X(T), X(T)\big\rangle + 2\int_{-\delta}^0 \big\langle G_{10}(\theta)^\top X(T+\theta), X(T)\big\rangle d\theta \\ &+ \int_{[-\delta,0]^2} \big\langle G_{11}(\theta,\theta')X(T+\theta), X(T+\theta')\big\rangle d\theta' d\theta \\ &+2\big\langle g_0, X(T)\big\rangle + 2\int_{-\delta}^0 \big\langle g_1(\theta), X(T+\theta)\big\rangle d\theta. \end{aligned} \tag{3.3.2}$$

The assumption (H2.2) becomes the following:

(H3.3) Let $T \geqslant \delta > 0$ be given.

(i) The coefficients of the state Eq. (3.3.1) satisfy the following assumptions:

$$\begin{aligned} &A(\cdot,\cdot) \in C([0,T+\delta]\times[-\delta,0];\mathbb{R}^{n\times n}), \\ &B_0(\cdot) \in C([0,T+\delta];\mathbb{R}^{n\times m}), \quad b(\cdot) \in L^1(0,T;\mathbb{R}^n). \end{aligned}$$

(ii) The weight coefficients of the cost functional (3.3.2) are all continuous, and $Q_{11}(t,\theta,\theta')^\top = Q_{11}(t,\theta',\theta)$, $G_{11}(\theta,\theta')^\top = G_{11}(\theta',\theta)$.

Hence when Problem (P) involves only the state delays not the control delays, it becomes the following problem.

Problem (P-NCD) For any $(s,x,\varphi(\cdot)) \in [0,T)\times M^2$, to find a $u^*(\cdot) \in L^2(s,T;\mathbb{R}^m)$ such that (3.3.1) is satisfied and

$$J(s,x,\varphi(\cdot);u^*(\cdot)) = \inf_{u(\cdot)\in L^2(s,T;\mathbb{R}^m)} J(s,x,\varphi(\cdot);u(\cdot)) := V(s,x,\varphi(\cdot)).$$

Similarly, in the special case when $b(\cdot)$, $q_0(\cdot)$, $q_1(\cdot,\cdot)$, $\rho_0(\cdot)$, g_0 and $g_1(\cdot)$ vanish, denote the corresponding delayed deterministic LQ problem, cost functional, and the value function by Problem (P_0-NCD), $J_0(s,x,\varphi(\cdot);u(\cdot))$ and $V_0(s,x,\varphi(\cdot))$, respectively.

As Sect. 2.2, we can lift the state Eq. (3.3.1) from $\mathbb{R}^n$ to M^2, and hence transforms the state equation with delays into one without delays. In this case, (2.1.29) becomes

$$\mathbf{X}(t) = \Phi(t,s)\xi + \int_s^t \Phi(t,r)\big[\tilde{B}(r)u(r) + \tilde{b}(r)\big]dr, \quad t \in [s,T], \tag{3.3.3}$$

where $\Phi(t,s)$ is defined as (2.1.10), $\widetilde{B}(r)$ defined by (2.1.13), is reduced to, $\tilde{B}(r)u : \mathbb{R}^m \to M^2$, $u \mapsto \begin{pmatrix} B_0(r)u \\ 0 \end{pmatrix}$, and $\tilde{b}(r) := \begin{pmatrix} b(r) \\ 0 \end{pmatrix}$, $\xi := \begin{pmatrix} x \\ \varphi \end{pmatrix}$. Meanwhile, (2.2.9) becomes

$$\begin{aligned} J(s,\xi;u(\cdot)) &:= J(s,x,\varphi(\cdot);u(\cdot)) \\ &= \int_s^T \Big[\big\langle \tilde{Q}(t)\mathbf{X}(t), \mathbf{X}(t)\big\rangle_{M^2} + 2\big\langle \tilde{S}_0(t)\mathbf{X}(t), u(t)\big\rangle \\ &\qquad + \big\langle \tilde{R}_{00}(t)u(t), u(t)\big\rangle + 2\big\langle \tilde{q}(t), \mathbf{X}(t)\big\rangle_{M^2} + 2\big\langle \tilde{\rho}_0(t), u(t)\big\rangle\Big]dt \\ &\quad + \big\langle \tilde{G}\mathbf{X}(T), \mathbf{X}(T)\big\rangle_{M^2} + 2\big\langle \tilde{g}, \mathbf{X}(T)\big\rangle_{M^2}. \end{aligned}$$

Here

$$\tilde{Q}(t) := \begin{bmatrix} \tilde{Q}_{00}(t) & \tilde{Q}_{01}(t) \\ \tilde{Q}_{10}(t) & \tilde{Q}_{11}(t) \end{bmatrix}, \quad \tilde{S}_0(t) := \begin{bmatrix} \tilde{S}_{00}(t) & \tilde{S}_{01}(t) \end{bmatrix}, \quad \tilde{R}_{00}(t) := R_{00}(t),$$

$$\tilde{G} := \begin{bmatrix} \tilde{G}_{00} & \tilde{G}_{01} \\ \tilde{G}_{10} & \tilde{G}_{11} \end{bmatrix}, \quad \tilde{q}(t) := \begin{bmatrix} \tilde{q}_0(t) \\ \tilde{q}_1(t) \end{bmatrix}, \quad \tilde{\rho}_0(t) := \rho_0(t), \quad \tilde{g} := \begin{bmatrix} \tilde{g}_0 \\ \tilde{g}_1 \end{bmatrix}.$$

Now Problem (EP) becomes the following problem.

Problem (EP-NCD) For any $(s,\xi) \in [0,T) \times M^2$, to find a $u^*(\cdot) \in L^2(s,T;\mathbb{R}^m)$ such that (3.3.3) is satisfied and

$$J(s,\xi;u^*(\cdot)) = \inf_{u(\cdot)\in L^2(s,T;\mathbb{R}^m)} J(s,\xi;u(\cdot)) := V(s,\xi).$$

Similarly, in the special case when $b(\cdot)$, $q_0(\cdot)$, $q_1(\cdot,\cdot)$, $\rho_0(\cdot)$, g_0 and $g_1(\cdot)$ vanish, denote the corresponding LQ problem, cost functional, and the value function by Problem (EP$_0$-NCD), $J_0(s,\xi;u(\cdot))$ and $V_0(s,\xi)$, respectively.

Now the definition of the closed-loop solvability for Problem (P-NCD) is simplified as follows.

Definition 3.3.1 Let

$$\begin{aligned} \Xi[s,T] := \Big\{ &\Theta(\cdot) : [s,T] \to \mathscr{L}(M^2,\mathbb{R}^m) \mid \Theta(\cdot) \text{ is strongly continuous} \\ &\text{and } \sup_{s\leqslant t\leqslant T} \|\Theta(t)\|_{\mathscr{L}(M^2,\mathbb{R}^m)} < \infty \Big\}, \end{aligned}$$

and $\mathcal{Q}[s,T] := \Xi[s,T]\times L^2(s,T;\mathbb{R}^m)$. Any pair $(\Theta(\cdot), v(\cdot)) \in \mathcal{Q}[s,T]$ is called a *closed-loop strategy of Problem (P-NCD)* on $[s,T]$. For any $(\Theta(\cdot), v(\cdot)) \in \mathcal{Q}[s,T]$ and $(x,\varphi) \in M^2$, let $\xi = \begin{pmatrix} x \\ \varphi \end{pmatrix}$, $\mathbf{X}(\cdot) \equiv \mathbf{X}(\cdot\,; s,\xi,\Theta(\cdot), v(\cdot))$ be the solution to the following equation:

$$\mathbf{X}(t) = \Phi(t,s)\xi + \int_s^t \Phi(t,r)\big[\tilde{B}(r)\big(\Theta(r)\mathbf{X}(r) + v(r)\big) + \tilde{b}(r)\big]dr,\ t \in [s,T],$$

and let

$$u(t) = \Theta(t)\mathbf{X}(t) + v(t), \quad t \in [s,T],$$

then $(\mathbf{X}(\cdot), u(\cdot))$ is called the *outcome pair* of $(\Theta(\cdot), v(\cdot))$ on $[s,T]$ corresponding to the initial trajectory $(x,\varphi(\cdot))$; $\mathbf{X}(\cdot)$, $u(\cdot)$ are called the corresponding *closed-loop state* and *closed-loop outcome control*, respectively.

Definition 3.3.2 A closed-loop strategy $(\Theta^*(\cdot), v^*(\cdot)) \in \mathcal{Q}[s,T]$ is said to be *optimal* on $[s,T]$ if

$$\begin{aligned} J\big(s,\xi;\Theta^*(\cdot)\mathbf{X}^*(\cdot)+v^*(\cdot)\big) &\leqslant J\big(s,\xi;\Theta(\cdot)\mathbf{X}(\cdot)+v(\cdot)\big), \\ &\forall\,(\Theta(\cdot), v(\cdot)) \in \mathcal{Q}[s,T],\ \forall\,\xi = \begin{pmatrix} x \\ \varphi \end{pmatrix} \in M^2, \end{aligned}$$

where $\mathbf{X}^*(\cdot)$, $\mathbf{X}(\cdot)$ are the closed-loop states corresponding to $(\Theta^*(\cdot), v^*(\cdot), x, \varphi)$, $(\Theta(\cdot), v(\cdot), x, \varphi)$, respectively. If an optimal closed-loop strategy (uniquely) exists on $[s,T]$, Problem (P-NCD) is said to be *(uniquely) closed-loop solvable on* $[s,T]$.

Similar to Theorem 3.1.1, we have the following results.

Lemma 3.3.3 *Let* (H3.3) *hold. For any given initial pair* $(s,x,\varphi(\cdot)) \in [0,T) \times M^2$, $u^*(\cdot)$ *is an open-loop optimal control of* Problem (P-NCD) *if and only if the following two conditions hold:*

(i) (Stationarity condition)

$$\tilde{S}_0(t)\mathbf{X}^*(t) + \tilde{R}_{00}(t)u^*(t) + \tilde{B}(t)^* p^*(t) + \tilde{\rho}_0(t) = 0, \ \text{a.e. } t \in [s,T],$$

where $(\mathbf{X}^*(\cdot), p^*(\cdot)) \in C([s,T];M^2) \times C([s,T];M^2)$ *satisfies the following integral equation with* $\xi := \begin{pmatrix} x \\ \varphi \end{pmatrix}$:

$$\begin{cases} \mathbf{X}^*(t) = \Phi(t,s)\xi + \displaystyle\int_s^t \Phi(t,r)\big[\tilde{B}(r)u^*(r) + \tilde{b}(r)\big]dr, \quad t \in [s,T], \\ p^*(t) = \Phi(T,t)^*\big[\tilde{G}\mathbf{X}^*(T) + \tilde{g}\big] + \displaystyle\int_t^T \Phi(r,t)^*\big[\tilde{Q}(r)\mathbf{X}^*(r) \\ \qquad + \tilde{S}_0(r)^* u^*(r) + \tilde{q}(r)\big]dr, \quad t \in [s,T]. \end{cases}$$

(ii) (Convexity condition)

$$\int_s^T \Big\{\langle \tilde{Q}(t)\mathbf{X}^0(t), \mathbf{X}^0(t)\rangle_{M^2} + 2\langle \tilde{S}_0(t)\mathbf{X}^0(t), u^0(t)\rangle$$
$$+\langle \tilde{R}_{00}(t)u^0(t), u^0(t)\rangle\Big\}dt + \langle \tilde{G}\mathbf{X}^0(T), \mathbf{X}^0(T)\rangle_{M^2} \geqslant 0,$$
$$\forall\, u^0(\cdot) \in L^2(s, T; \mathbb{R}^m),$$

where $\mathbf{X}^0(\cdot)$ *is the solution to the following integral equation:*

$$\mathbf{X}^0(t) = \int_s^t \Phi(t, r)\tilde{B}(r)u^0(r)dr, \quad t \in [s, T].$$

By Lemma 3.3.3, we can derive the following result, which plays an important role in the study of closed-loop solvability for Problem (P-NCD).

Proposition 3.3.4 *Let* (H3.3) *hold. Suppose* $(\Theta^*(\cdot), v^*(\cdot))$ *is the optimal closed-loop strategy of* Problem (P-NCD) *on* $[s, T]$*, then* $(\Theta^*(\cdot), 0)$ *is the optimal closed-loop strategy of* Problem (P_0-NCD) *on* $[s, T]$.

Proof Suppose $(\Theta^*(\cdot), v^*(\cdot))$ is the optimal closed-loop strategy of Problem (P-NCD), then $v^*(\cdot)$ is the open-loop optimal control for the optimal control problem with the following state equation

$$\mathbf{X}(t) = \Phi(t, s)\xi + \int_s^t \Phi(t, r)\big[\tilde{B}(r)(\Theta^*(r)\mathbf{X}(r) + v(r)) + \tilde{b}(r)\big]dr, \quad t \in [s, T],$$

and the following cost functional

$$\tilde{J}(s, \xi; v(\cdot)) := \int_s^T \Big\{\langle[\tilde{Q}(t) + \Theta^*(t)^*\tilde{S}_0(t) + \tilde{S}_0(t)^*\Theta^*(t)$$
$$+\Theta^*(t)^*\tilde{R}_{00}(t)\Theta^*(t)]\mathbf{X}(t), \mathbf{X}(t)\rangle_{M^2}$$
$$+2\langle[\tilde{S}_0(t) + \tilde{R}_{00}(t)\Theta^*(t)]\mathbf{X}(t), v(t)\rangle$$
$$+\langle \tilde{R}_{00}(t)v(t), v(t)\rangle + 2\langle \tilde{q}(t) + \Theta^*(t)^*\tilde{\rho}_0(t), \mathbf{X}(t)\rangle_{M^2}$$
$$+2\langle \tilde{\rho}_0(t), v(t)\rangle\Big\}dt + \langle \tilde{G}\mathbf{X}(T), \mathbf{X}(T)\rangle_{M^2} + 2\langle \tilde{g}, \mathbf{X}(T)\rangle_{M^2},$$

where $\Theta^*(t)^*$ is the adjoint operator of $\Theta^*(t)$. Denote this problem by Problem ($\tilde{\text{P}}$-NCD). Similar to Lemma 3.3.3, we derive

$$[\tilde{S}_0(t) + \tilde{R}_{00}(t)\Theta^*(t)]\mathbf{X}^*(t) + \tilde{R}_{00}(t)v^*(t)$$
$$+\tilde{B}(t)^*p^*(t) + \tilde{\rho}_0(t) = 0, \quad \text{a.e. } t \in [s, T],$$

where $(\mathbf{X}^*(\cdot), p^*(\cdot))$ satisfies the following integral equation:

$$\begin{cases} \mathbf{X}^*(t) = \Phi_\Theta(t,s)\xi + \int_s^t \Phi_\Theta(t,r)\big[\tilde{B}(r)v^*(r) + \tilde{b}(r)\big]dr, \ t \in [s,T], \\ p^*(t) = \Phi_\Theta(T,t)^*\big[\tilde{G}\mathbf{X}^*(T) + \tilde{g}\big] + \int_t^T \Phi_\Theta(r,t)^*\Big\{\big[\Theta^*(r)^*\tilde{S}_0(r) \\ \qquad +\tilde{S}_0(r)^*\Theta^*(r) + \Theta^*(r)^*\tilde{R}_{00}(r)\Theta^*(r) + \tilde{Q}(r)\big]\mathbf{X}^*(r) \\ \qquad +\big[\tilde{S}_0(r)^* + \Theta^*(r)^*\tilde{R}_{00}(r)\big]v^*(r) \\ \qquad +\tilde{q}(r) + \Theta^*(r)^*\tilde{\rho}_0(r)\Big\}dr, \ t \in [s,T], \end{cases}$$

with

$$\Phi_\Theta(t,s)\xi = \Phi(t,s)\xi + \int_s^t \Phi(t,r)\tilde{B}(r)\Theta^*(r)\Phi_\Theta(r,s)\xi dr, \quad t \in [s,T].$$

Since the above admits a solution for each $\xi \in M^2$, and $(\Theta^*(\cdot), v^*(\cdot))$ is independent of ξ, by subtracting solutions corresponding ξ and 0, the later from the former, we see that for any $\xi \in M^2$,

$$\big[\tilde{S}_0(t) + \tilde{R}_{00}(t)\Theta^*(t)\big]\mathbf{X}(t) + \tilde{B}(t)^*p(t) = 0, \quad \text{a.e. } t \in [s,T],$$

where

$$\begin{cases} \mathbf{X}(t) = \Phi_\Theta(t,s)\xi, \quad t \in [s,T], \\ p(t) = \Phi_\Theta(T,t)^*\tilde{G}\mathbf{X}(T) + \int_t^T \Phi_\Theta(r,t)^*\big[\tilde{Q}(r) + \Theta^*(r)^*\tilde{S}_0(r) \\ \qquad +\tilde{S}_0(r)^*\Theta^*(r) + \Theta^*(r)^*\tilde{R}_{00}(r)\Theta^*(r)\big]\mathbf{X}(r)dr, \quad t \in [s,T]. \end{cases}$$

Again similar to Lemma 3.3.3, for any $\xi \in M^2$, 0 is the open-loop optimal control of Problem ($\tilde{\text{P}}_0$-NCD). Therefore $(\Theta^*(\cdot), 0)$ is the optimal closed-loop strategy of Problem (P_0-NCD). This completes the proof. □

Next we shall study the necessary conditions of the closed-loop solvability for Problem (P-NCD). We borrow some ideas from Lü [9]. Consider the following integral equation:

$$\begin{aligned} P(t)\xi &= \Phi_\Theta(T,t)^*\tilde{G}\Phi_\Theta(T,t)\xi + \int_t^T \Phi_\Theta(r,t)^*\big[\tilde{Q}(r) + \Theta^*(r)^*\tilde{S}_0(r) \\ &\quad +\tilde{S}_0(r)^*\Theta^*(r) + \Theta^*(r)^*\tilde{R}_{00}(r)\Theta^*(r)\big]\Phi_\Theta(r,t)\xi dr, \ t \in [s,T], \end{aligned} \tag{3.3.4}$$

where $\Phi_\Theta(\cdot,\cdot)$ satisfies

$$\Phi_\Theta(t,s)\xi = \Phi(t,s)\xi + \int_s^t \Phi(t,r)\tilde{B}(r)\Theta^*(r)\Phi_\Theta(r,s)\xi dr, \quad t \in [s,T], \tag{3.3.5}$$

and $\Theta^*(r)^*$ is the adjoint operator of $\Theta^*(r)$. Since $\Theta^*(\cdot) \in \Xi[s, T]$ and $\Phi(\cdot, \cdot)$ is a strongly continuous mild evolution operator, (3.3.5) has a unique solution $\Phi_\Theta(\cdot, \cdot)$ on $\Delta_*[0, T]$ in the class of strongly continuous bounded linear operators on M^2, hence (3.3.4) is well-defined. Moreover $\Phi_\Theta(\cdot, \cdot)$ is a mild evolution operator and also satisfies the following integral equation:

$$\Phi_\Theta(t, s)\xi = \Phi(t, s)\xi + \int_s^t \Phi_\Theta(t, r)\tilde{B}(r)\Theta^*(r)\Phi(r, s)\xi dr, \quad t \in [s, T].$$

Then we can derive the following necessary conditions of the closed-loop solvability for Problem (P_0-NCD).

Theorem 3.3.5 *Let* (H3.3) *hold. Suppose* $(\Theta^*(\cdot), 0)$ *is the optimal closed-loop strategy of* Problem (P_0-NCD) *on* $[s, T]$, *then*

$$\tilde{R}_{00}(t) \geqslant 0, \ \left[\tilde{R}_{00}(t)\Theta^*(t) + \tilde{B}(t)^* P(t) + \tilde{S}_0(t)\right]\Phi_\Theta(t, s) = 0, \ \text{a.e. } t \in [s, T],$$

where $P(\cdot)$ *satisfies (3.3.4).*

Proof In the following proof, we simply write $\langle \cdot , \cdot \rangle_{M^2}$, $\langle \cdot , \cdot \rangle_{L^2}$ and $\langle \cdot , \cdot \rangle_Z$ as $\langle \cdot , \cdot \rangle$ if no ambiguity exists. First we aim to prove that

$$\begin{aligned} &J_0(s, \xi; \Theta^*(\cdot)z(\cdot) + v(\cdot)) \\ &= \langle P(s)\xi, \xi\rangle + \int_s^T \Big\{\langle \tilde{R}_{00}(t)v(t), v(t)\rangle + 2\langle\big[\tilde{R}_{00}(t)\Theta^*(t) \\ &\qquad + \tilde{B}(t)^* P(t) + \tilde{S}_0(t)\big]z(t), v(t)\rangle\Big\}dt, \end{aligned} \tag{3.3.6}$$

where

$$\begin{aligned} z(t) &= \Phi(t, s)\xi + \int_s^t \Phi(t, r)\tilde{B}(r)\big[\Theta^*(r)z(r) + v(r)\big]dr \\ &= \Phi_\Theta(t, s)\xi + \int_s^t \Phi_\Theta(t, r)\tilde{B}(r)v(r)dr, \quad t \in [s, T]. \end{aligned} \tag{3.3.7}$$

By the definition of J_0, we have

$$\begin{aligned} &J_0(s, \xi; \Theta^*(\cdot)z(\cdot) + v(\cdot)) \\ &= \int_s^T \Big\{\langle\big[\tilde{Q}(t) + \Theta^*(t)^*\tilde{S}_0(t) + \tilde{S}_0(t)^*\Theta^*(t) + \Theta^*(t)^*\tilde{R}_{00}(t)\Theta^*(t)\big]z(t), z(t)\rangle \\ &\quad + 2\langle\tilde{S}_0(t)z(t), v(t)\rangle + 2\langle\tilde{R}_{00}(t)\Theta^*(t)z(t), v(t)\rangle \\ &\quad + \langle\tilde{R}_{00}(t)v(t), v(t)\rangle\Big\}dt + \langle\tilde{G}z(T), z(T)\rangle, \end{aligned}$$

which and (3.3.6) imply that we only need to prove that

$$\int_s^T \big\langle\big[\tilde{Q}(t) + \Theta^*(t)^*\tilde{S}_0(t) + \tilde{S}_0(t)^*\Theta^*(t) + \Theta^*(t)^*\tilde{R}_{00}(t)\Theta^*(t)\big]z(t), z(t)\big\rangle dt$$
$$+\big\langle\tilde{G}z(T), z(T)\big\rangle = \big\langle P(s)\xi, \xi\big\rangle + 2\int_s^T \big\langle\tilde{B}(t)^*P(t)z(t), v(t)\big\rangle dt. \tag{3.3.8}$$

Denote

$$f(t) := \tilde{Q}(t) + \Theta^*(t)^*\tilde{S}_0(t) + \tilde{S}_0(t)^*\Theta^*(t) + \Theta^*(t)^*\tilde{R}_{00}(t)\Theta^*(t).$$

By (3.3.7), we obtain

$$\begin{aligned}
&\text{L. H. S. of (3.3.8)} = \int_s^T \big\langle f(t)\Phi_\Theta(t,s)\xi, \Phi_\Theta(t,s)\xi\big\rangle dt\\
&+2\int_s^T \Big\langle f(t)\Phi_\Theta(t,s)\xi, \int_s^t \Phi_\Theta(t,r)\tilde{B}(r)v(r)dr\Big\rangle dt\\
&+\int_s^T \Big\langle f(t)\int_s^t \Phi_\Theta(t,r)\tilde{B}(r)v(r)dr, \int_s^t \Phi_\Theta(t,r)\tilde{B}(r)v(r)dr\Big\rangle dt\\
&+\big\langle\tilde{G}\Phi_\Theta(T,s)\xi, \Phi_\Theta(T,s)\xi\big\rangle + 2\Big\langle\tilde{G}\Phi_\Theta(T,s)\xi, \int_s^T \Phi_\Theta(T,r)\tilde{B}(r)v(r)dr\Big\rangle\\
&+\Big\langle\tilde{G}\int_s^T \Phi_\Theta(T,r)\tilde{B}(r)v(r)dr, \int_s^T \Phi_\Theta(T,r)\tilde{B}(r)v(r)dr\Big\rangle.
\end{aligned} \tag{3.3.9}$$

By (3.3.4) and (3.3.5), we derive

$$\begin{aligned}
&2\int_s^T \big\langle\tilde{B}(t)^*P(t)z(t), v(t)\big\rangle dt = 2\int_s^T \big\langle P(t)z(t), \tilde{B}(t)v(t)\big\rangle dt\\
&= 2\int_s^T \Big\langle\tilde{B}(t)v(t), \Phi_\Theta(T,t)^*\tilde{G}\Phi_\Theta(T,s)\xi\\
&\qquad +\Phi_\Theta(T,t)^*\tilde{G}\int_s^t \Phi_\Theta(T,r)\tilde{B}(r)v(r)dr\Big\rangle dt\\
&\quad +2\int_s^T \Big\langle\tilde{B}(t)v(t), \int_t^T \Phi_\Theta(r,t)^*f(r)\Phi_\Theta(r,s)\xi dr\Big\rangle dt\\
&\quad +2\int_s^T \Big\langle\tilde{B}(t)v(t), \int_t^T \Phi_\Theta(r,t)^*f(r)\int_s^t \Phi_\Theta(r,\alpha)\tilde{B}(\alpha)v(\alpha)d\alpha dr\Big\rangle dt.
\end{aligned} \tag{3.3.10}$$

Notice that

$$
\begin{aligned}
&2\int_s^T \Big\langle \tilde{B}(t)v(t), \int_t^T \Phi_\Theta(r,t)^* f(r)\int_s^t \Phi_\Theta(r,\alpha)\tilde{B}(\alpha)v(\alpha)d\alpha dr\Big\rangle dt\\
&=\int_s^T\int_s^r\int_s^t \Big\langle \Phi_\Theta(r,t)\tilde{B}(t)v(t), f(r)\Phi_\Theta(r,\alpha)\tilde{B}(\alpha)v(\alpha)\Big\rangle d\alpha dt dr\\
&\quad+\int_s^T\int_s^r\int_\alpha^r \Big\langle \Phi_\Theta(r,t)\tilde{B}(t)v(t), f(r)\Phi_\Theta(r,\alpha)\tilde{B}(\alpha)v(\alpha)\Big\rangle dt d\alpha dr\\
&=\int_s^T\int_s^r \Big\langle \Phi_\Theta(r,\alpha)\tilde{B}(\alpha)v(\alpha), f(r)\int_s^\alpha \Phi_\Theta(r,t)\tilde{B}(t)v(t)dt\Big\rangle d\alpha dr\\
&\quad+\int_s^T\int_s^r \Big\langle \int_\alpha^r \Phi_\Theta(r,t)\tilde{B}(t)v(t)dt, f(r)\Phi_\Theta(r,\alpha)\tilde{B}(\alpha)v(\alpha)\Big\rangle d\alpha dr\\
&=\int_s^T \Big\langle \int_s^r \Phi_\Theta(r,t)\tilde{B}(t)v(t)dt, f(r)\int_s^r \Phi_\Theta(r,\alpha)\tilde{B}(\alpha)v(\alpha)d\alpha\Big\rangle dr.
\end{aligned}
\tag{3.3.11}
$$

Combining (3.3.9), (3.3.10), (3.3.11) and (3.3.4), (3.3.8) is proved and thus (3.3.6) holds. Since $(\Theta^*(\cdot), 0)$ is the optimal closed-loop strategy of Problem (P_0-NCD),

$$
J_0(s,\xi;\Theta^*(\cdot)z^*(\cdot)) \leqslant J_0(s,\xi;\Theta^*(\cdot)z(\cdot)+v(\cdot)),
$$

which and (3.3.6) imply that

$$
\begin{aligned}
&\int_s^T \Big\{\big\langle \tilde{R}_{00}(t)v(t), v(t)\big\rangle + 2\big\langle\big[\tilde{R}_{00}(t)\Theta^*(t)+\tilde{B}(t)^*P(t)\\
&\quad+\tilde{S}_0(t)\big]z(t), v(t)\big\rangle\Big\}dt \geqslant 0, \quad \forall\, v(\cdot)\in L^2(s,T;\mathbb{R}^m),\ \xi\in M^2,
\end{aligned}
\tag{3.3.12}
$$

where $z(\cdot)$ satisfies (3.3.7). Next we aim to prove that

$$
\tilde{R}_{00}(t) \geqslant 0, \quad \text{a.e. } t\in[s,T]. \tag{3.3.13}
$$

Suppose there exists $\Omega_0 \subseteq [s,T]$ and $|\Omega_0| > 0$ such that $\tilde{R}_{00}(t) < 0$ on $t \in \Omega_0$, that is, there exists $\beta > 0$ such that $\tilde{R}_{00}(t) \leqslant -\beta I$ on Ω_0. Without loss of generality, assume $|\Omega_0| > \frac{1}{k}$. Choose a sequence of Borel measurable sets $\{\Omega_i\}$ such that $\Omega_i \subseteq \Omega_0$ and $|\Omega_i| = \frac{1}{k+i}$ for $i = 1, 2, \cdots$. Choose $\xi = 0$ and $v_i(t) = \big(i, 0, \cdots, 0\big)^\top_{\mathbb{R}^{1\times m}} \mathbf{1}_{\Omega_i}(t)$, let $z_i(\cdot)$ be the solution to (3.3.7) with $v_i(\cdot)$, then there exists a constant $K > 0$ independent of i such that

$$
\sup_{s\leqslant t\leqslant T} \|z_i(t)\| \leqslant K.
$$

It follows that

$$\left|\int_s^T \big\langle\big[\tilde{R}_{00}(t)\Theta^*(t)+\tilde{B}(t)^*P(t)+\tilde{S}_0(t)\big]z_i(t), v_i(t)\big\rangle dt\right| \leqslant \frac{Ki}{k+i},$$

thus we get

$$\varlimsup_{i\to\infty}\frac{1}{i}\int_s^T \Big\{\big\langle\tilde{R}_{00}(t)v_i(t), v_i(t)\big\rangle + 2\big\langle\big[\tilde{R}_{00}(t)\Theta^*(t)+\tilde{B}(t)^*P(t) +\tilde{S}_0(t)\big]z_i(t), v_i(t)\big\rangle\Big\}dt \leqslant -\beta,$$

which contradicts (3.3.12). Hence (3.3.13) is proved.

Finally we aim to prove that

$$\big[\tilde{R}_{00}(t)\Theta^*(t)+\tilde{B}(t)^*P(t)+\tilde{S}_0(t)\big]\Phi_\Theta(t,s)=0, \quad \text{a.e. } t\in[s,T]. \tag{3.3.14}$$

Choose $\xi\in M^2$ and $v_j(t)=\frac{1}{j}v(t)$, $v(\cdot)\in L^2(s,T;\mathbb{R}^m)$, then

$$z_j(t)=\Phi_\Theta(t,s)\xi+\frac{1}{j}\int_s^t \Phi_\Theta(t,r)\tilde{B}(r)v(r)dr, \quad t\in[s,T]. \tag{3.3.15}$$

Apparently

$$\sup_{s\leqslant t\leqslant T}\|z_j(t)-\Phi_\Theta(t,s)\xi\|\to 0, \quad \text{as } j\to\infty.$$

From (3.3.12) we have

$$\varlimsup_{j\to\infty} j\int_s^T \Big\{\big\langle\tilde{R}_{00}(t)v_j(t), v_j(t)\big\rangle + 2\big\langle\big[\tilde{R}_{00}(t)\Theta^*(t)+\tilde{B}(t)^*P(t) +\tilde{S}_0(t)\big]z_j(t), v_j(t)\big\rangle\Big\}dt \geqslant 0, \quad \forall\, v(\cdot)\in L^2(s,T;\mathbb{R}^m),\ \xi\in M^2.$$

It follows that

$$\varlimsup_{j\to\infty}\int_s^T \big\langle\big[\tilde{R}_{00}(t)\Theta^*(t)+\tilde{B}(t)^*P(t)+\tilde{S}_0(t)\big]z_j(t), v(t)\big\rangle dt \geqslant 0, \quad \forall\, v(\cdot)\in L^2(s,T;\mathbb{R}^m), \quad \xi\in M^2,$$

which and (3.3.15) imply that

$$\int_s^T \big\langle\big[\tilde{R}_{00}(t)\Theta^*(t)+\tilde{B}(t)^*P(t)+\tilde{S}_0(t)\big]\Phi_\Theta(t,s)\xi, v(t)\big\rangle dt \geqslant 0, \quad \forall\, v(\cdot)\in L^2(s,T;\mathbb{R}^m).$$

Choose $v(t) = v\mathbf{1}_{[\tau,\tau+\varepsilon]}(t)$, $v \in \mathbb{R}^m$, $\tau \in [s, T)$, then

$$\begin{aligned}&\big\langle\big[\tilde{R}_{00}(\tau)\Theta^*(\tau) + \tilde{B}(\tau)^* P(\tau) + \tilde{S}_0(\tau)\big]\Phi_\Theta(\tau, s)\xi, v\big\rangle = 0,\\ &\qquad\qquad\qquad\qquad\qquad\text{a.e. } \tau \in [s, T],\ \forall\, v \in \mathbb{R}^m,\ \xi \in M^2.\end{aligned}$$

By the arbitrariness of ξ and v, (3.3.14) is proved. Hence the proof is completed. □

Now we give the necessary conditions of the closed-loop solvability for Problem (P-NCD).

Theorem 3.3.6 *Let* (H3.3) *hold. Suppose* $(\Theta^*(\cdot), v^*(\cdot))$ *is the optimal closed-loop strategy of* Problem (P-NCD) *on* $[s, T]$, *then*

$$\tilde{R}_{00}(t) \geqslant 0, \qquad \big[\tilde{R}_{00}(t)\Theta^*(t) + \tilde{B}(t)^* P(t) + \tilde{S}_0(t)\big]\Phi_\Theta(t, s) = 0, \tag{3.3.16}$$

$$\begin{aligned}&\big[\tilde{R}_{00}(t)\Theta^*(t) + \tilde{B}(t)^* P(t) + \tilde{S}_0(t)\big]\int_s^t \Phi_\Theta(t, r)\big[\tilde{B}(r)v^*(r) + \tilde{b}(r)\big]dr\\ &\quad+\big[\tilde{R}_{00}(t)v^*(t) + \tilde{B}(t)^*\eta(t) + \tilde{\rho}_0(t)\big] = 0, \quad \text{a.e. } t \in [s, T],\end{aligned} \tag{3.3.17}$$

where for any $\xi \in M^2$, $P(\cdot)$ *satisfies the following integral Riccati equation:*

$$\begin{aligned}P(t)\xi &= \Phi_\Theta(T, t)^*\tilde{G}\Phi_\Theta(T, t)\xi + \int_t^T \Phi_\Theta(r, t)^*\big[\tilde{Q}(r) + \Theta^*(r)^*\tilde{S}_0(r)\\ &\quad+\tilde{S}_0(r)^*\Theta^*(r) + \Theta^*(r)^*\tilde{R}_{00}(r)\Theta^*(r)\big]\Phi_\Theta(r, t)\xi dr,\ t \in [s, T],\end{aligned} \tag{3.3.18}$$

with $\Phi_\Theta(\cdot, \cdot)$ *satisfying the following integral equation:*

$$\Phi_\Theta(t, s)\xi = \Phi(t, s)\xi + \int_s^t \Phi(t, r)\tilde{B}(r)\Theta^*(r)\Phi_\Theta(r, s)\xi dr, \quad t \in [s, T],$$

and $\Theta^*(r)^*$ *being the adjoint operator of* $\Theta^*(r)$, *and* $\eta(\cdot)$ *satisfies the following integral equation:*

$$\begin{aligned}\eta(t) &= \Phi_\Theta(T, t)^*\tilde{g} + \int_t^T \Phi_\Theta(r, t)^*\Big\{\tilde{q}(r) + P(r)\tilde{b}(r) + \Theta^*(r)^*\tilde{\rho}_0(r)\\ &\quad+\big[\tilde{S}_0(r) + \tilde{R}_{00}(r)\Theta^*(r) + \tilde{B}(r)^*P(r)\big]^*v^*(r)\Big\}dr,\ t \in [s, T].\end{aligned}$$

Proof By Proposition 3.3.4 and Theorem 3.3.5, (3.3.16) is proved, thus we only need to prove (3.3.17). As in the proof of Proposition 3.3.4, since $(\Theta^*(\cdot), v^*(\cdot))$ is the optimal closed-loop strategy of Problem (P-NCD), $v^*(\cdot)$ is the open-loop optimal control of Problem (P̃-NCD). Hence we derive

$$\begin{aligned}&\big[\tilde{S}_0(t)+\tilde{R}_{00}(t)\Theta^*(t)\big]\mathbf{X}^*(t)+\tilde{R}_{00}(t)v^*(t)\\&\quad+\tilde{B}(t)^*p^*(t)+\tilde{\rho}_0(t)=0,\quad \text{a.e. } t\in[s,T],\end{aligned}\tag{3.3.19}$$

where $(\mathbf{X}^*(\cdot), p^*(\cdot))$ satisfies the following integral equation:

$$\begin{cases}\mathbf{X}^*(t)=\Phi_\Theta(t,s)\xi+\displaystyle\int_s^t\Phi_\Theta(t,r)\big[\tilde{B}(r)v^*(r)+\tilde{b}(r)\big]dr,\ t\in[s,T],\\ p^*(t)=\Phi_\Theta(T,t)^*\big[\tilde{G}\mathbf{X}^*(T)+\tilde{g}\big]+\displaystyle\int_t^T\Phi_\Theta(r,t)^*\Big\{\big[\tilde{Q}(r)+\Theta^*(r)^*\tilde{S}_0(r)\\ \qquad+\tilde{S}_0(r)^*\Theta^*(r)+\Theta^*(r)^*\tilde{R}_{00}(r)\Theta^*(r)\big]\mathbf{X}^*(r)+\big[\tilde{S}_0(r)^*\\ \qquad+\Theta^*(r)^*\tilde{R}_{00}(r)\big]v^*(r)+\tilde{q}(r)+\Theta^*(r)^*\tilde{\rho}_0(r)\Big\}dr,\ t\in[s,T].\end{cases}\tag{3.3.20}$$

Now we claim that

$$p^*(t)=P(t)\mathbf{X}^*(t)+\eta(t),\ t\in[s,T].\tag{3.3.21}$$

Then by (3.3.19), we have

$$\begin{aligned}&\big[\tilde{S}_0(t)+\tilde{R}_{00}(t)\Theta^*(t)+\tilde{B}(t)^*P(t)\big]\mathbf{X}^*(t)\\&+\tilde{B}(t)^*\eta(t)+\tilde{R}_{00}(t)v^*(t)+\tilde{\rho}_0(t)=0,\\&\qquad\qquad \text{a.e. } t\in[s,T],\ \forall\,\xi\in M^2.\end{aligned}$$

Noting

$$\mathbf{X}^*(t)=\Phi_\Theta(t,s)\xi+\int_s^t\Phi_\Theta(t,r)\big[\tilde{B}(r)v^*(r)+\tilde{b}(r)\big]dr,\quad t\in[s,T],$$

and by the arbitrariness of ξ, we derive (3.3.17). Next we only need to prove (3.3.21). Using (3.3.18), (3.3.20), we get

$$\begin{aligned}p^*(t)-P(t)\mathbf{X}^*(t)&=\Phi_\Theta(T,t)^*\big[\tilde{G}\mathbf{X}^*(T)+\tilde{g}-\tilde{G}\Phi_\Theta(T,t)\mathbf{X}^*(t)\big]\\&+\int_t^T\Phi_\Theta(r,t)^*\Big\{\big[\tilde{Q}(r)+\Theta^*(r)^*\tilde{S}_0(r)+\tilde{S}_0(r)^*\Theta^*(r)\\&\quad+\Theta^*(r)^*\tilde{R}_{00}(r)\Theta^*(r)\big]\big[\mathbf{X}^*(r)-\Phi_\Theta(r,t)\mathbf{X}^*(t)\big]\\&\quad+\big[\tilde{S}_0(r)^*+\Theta^*(r)^*\tilde{R}_{00}(r)\big]v^*(r)+\tilde{q}(r)+\Theta^*(r)^*\tilde{\rho}_0(r)\Big\}dr.\end{aligned}$$

Noting

$$\mathbf{X}^*(T) = \Phi_\Theta(T,t)\mathbf{X}^*(t) + \int_t^T \Phi_\Theta(T,r)\big[\tilde{B}(r)v^*(r) + \tilde{b}(r)\big]dr,$$

we have

$$\begin{aligned}
&p^*(t) - P(t)\mathbf{X}^*(t)\\
&= \Phi_\Theta(T,t)^*\tilde{g} + \int_t^T \Phi_\Theta(r,t)^*\Big\{\Phi_\Theta(T,r)^*\tilde{G}\Phi_\Theta(T,r)\big[\tilde{B}(r)v^*(r) + \tilde{b}(r)\big]\\
&\quad +\big[\tilde{Q}(r) + \Theta^*(r)^*\tilde{S}_0(r) + \tilde{S}_0(r)^*\Theta^*(r) + \Theta^*(r)^*\tilde{R}_{00}(r)\Theta^*(r)\big]\\
&\quad \times \int_t^r \Phi_\Theta(r,\alpha)\big[\tilde{B}(\alpha)v^*(\alpha) + \tilde{b}(\alpha)\big]d\alpha\\
&\quad +\big[\tilde{S}_0(r)^* + \Theta^*(r)^*\tilde{R}_{00}(r)\big]v^*(r) + \tilde{q}(r) + \Theta^*(r)^*\tilde{\rho}_0(r)\Big\}dr\\
&= \Phi_\Theta(T,t)^*\tilde{g} + \int_t^T \Phi_\Theta(r,t)^*\Big\{P(r)\big[\tilde{B}(r)v^*(r) + \tilde{b}(r)\big]\\
&\quad +\big[\tilde{S}_0(r)^* + \Theta^*(r)^*\tilde{R}_{00}(r)\big]v^*(r) + \tilde{q}(r) + \Theta^*(r)^*\tilde{\rho}_0(r)\Big\}dr = \eta(t),
\end{aligned}$$

which implies (3.3.21). This completes the proof of Theorem 3.3.6. □

Now we give the sufficient conditions of the closed-loop solvability for Problem (P-NCD).

Theorem 3.3.7 *Let* (H3.3) *hold. Suppose* $\tilde{R}_{00} \geqslant 0$, $\mathcal{R}(\tilde{B}^*P + \tilde{S}_0) \subseteq \mathcal{R}(\tilde{R}_{00})$ *and* $\mathcal{R}(\tilde{B}^*\eta + \tilde{\rho}_0) \subseteq \mathcal{R}(\tilde{R}_{00})$, *where for any* $\xi \in M^2$, $P(\cdot)$ *satisfies the following* integral operator-valued Riccati equations *(IOREs, for short):*

$$\begin{aligned}
&(a)\ P(t)\xi = \Phi(T,t)^*\tilde{G}\Phi(T,t)\xi + \int_t^T \Phi(r,t)^*\Big\{\tilde{Q}(r) - \Big(\tilde{R}_{00}(r)^\dagger\big[\tilde{B}(r)^*P(r)\\
&\qquad + \tilde{S}_0(r)\big]\Big)^*\big[\tilde{B}(r)^*P(r) + \tilde{S}_0(r)\big]\Big\}\Phi(r,t)\xi dr,\ t \in [s,T],\\
&(b)\ P(t)\xi = \Phi_\Theta(T,t)^*\tilde{G}\Phi_\Theta(T,t)\xi + \int_t^T \Phi_\Theta(r,t)^*\big[\tilde{Q}(r) + \Theta^*(r)^*\tilde{S}_0(r)\\
&\qquad +\tilde{S}_0(r)^*\Theta^*(r) + \Theta^*(r)^*\tilde{R}_{00}(r)\Theta^*(r)\big]\Phi_\Theta(r,t)\xi dr, t \in [s,T],\\
&(c)\ P(t)\xi = \Phi(T,t)^*\tilde{G}\Phi_\Theta(T,t)\xi + \int_t^T \Phi(r,t)^*\Big\{\tilde{Q}(r) - \tilde{S}_0(r)^*\tilde{R}_{00}(r)^\dagger\\
&\qquad \times\big[\tilde{B}(r)^*P(r) + \tilde{S}_0(r)\big]\Big\}\Phi_\Theta(r,t)\xi dr,\ t \in [s,T],
\end{aligned} \tag{3.3.22}$$

with $\Theta^*(r)^*$ being the adjoint operator of $\Theta^*(r)$, and $\eta(\cdot)$ satisfies the following backward integral evolution equations *(BIEEs, for short) :*

$$\begin{aligned}
&(a)\quad \eta(t)=\Phi_\Theta(T,t)^*\tilde g+\int_t^T\Phi_\Theta(r,t)^*\big[\tilde q(r)+P(r)\tilde b(r)\\
&\qquad\qquad +\Theta^*(r)^*\tilde\rho_0(r)\big]dr,\quad t\in[s,T],\\
&(b)\quad \eta(t)=\Phi(T,t)^*\tilde g+\int_t^T\Phi(r,t)^*\big[P(r)\tilde b(r)+\Theta^*(r)^*\tilde\rho_0(r)\\
&\qquad\qquad +\Theta^*(r)^*\tilde B(r)^*\eta(r)+\tilde q(r)\big]dr,\quad t\in[s,T],
\end{aligned}\tag{3.3.23}$$

with $\Phi(\cdot,\cdot)$ being the solution to the following integral equation:

$$\begin{aligned}
&(a)\ \Phi_\Theta(t,s)\xi=\Phi(t,s)\xi+\int_s^t\Phi(t,r)\tilde B(r)\Theta^*(r)\Phi_\Theta(r,s)\xi dr,\ t\in[s,T],\\
&(b)\ \Phi_\Theta(t,s)\xi=\Phi(t,s)\xi+\int_s^t\Phi_\Theta(t,r)\tilde B(r)\Theta^*(r)\Phi(r,s)\xi dr,\ t\in[s,T].
\end{aligned}\tag{3.3.24}$$

Moreover, in the above equations,

$$\begin{aligned}
&(a)\quad \Theta^*(t)=-\tilde R_{00}(t)^\dagger\big[\tilde B(t)^*P(t)+\tilde S_0(t)\big]\\
&\qquad\qquad +\big[I-\tilde R_{00}(t)^\dagger\tilde R_{00}(t)\big]\theta(t),\quad \text{a.e. } t\in[s,T],\\
&(b)\quad v^*(t)=-\tilde R_{00}(t)^\dagger\big[\tilde B(t)^*\eta(t)+\tilde\rho_0(t)\big]\\
&\qquad\qquad +\big[I-\tilde R_{00}(t)^\dagger\tilde R_{00}(t)\big]\varsigma(t),\quad \text{a.e. } t\in[s,T],
\end{aligned}\tag{3.3.25}$$

for any $\theta(\cdot)\in\Xi(s,T)$ and any $\varsigma(\cdot)\in L^2(s,T;\mathbb{R}^m)$. Suppose $(\Theta^*(\cdot),v^*(\cdot))\in\mathcal{Q}[s,T]$, then it is the optimal closed-loop strategy of Problem (P-NCD), and the value function is

$$\begin{aligned}
V(s,\xi)=&\int_s^T\Big\{2\langle\eta(t),\tilde b(t)\rangle_{M^2}-\langle\tilde R_{00}(t)^\dagger\big[\tilde B(t)^*\eta(t)+\tilde\rho_0(t)\big],\\
&\tilde B(t)^*\eta(t)+\tilde\rho_0(t)\rangle\Big\}dt+\langle P(s)\xi,\xi\rangle_{M^2}+2\langle\eta(s),\xi\rangle_{M^2}.
\end{aligned}\tag{3.3.26}$$

Proof In the following proof, we simply write $\langle\cdot,\cdot\rangle_{M^2}$, $\langle\cdot,\cdot\rangle_{L^2}$ and $\langle\cdot,\cdot\rangle_Z$ as $\langle\cdot,\cdot\rangle$ if no ambiguity exists. Noting $\mathcal{R}(\tilde B^*P+\tilde S_0)\subseteq\mathcal{R}(\tilde R_{00})$, $\mathcal{R}(\tilde B^*\eta+\tilde\rho_0)\subseteq\mathcal{R}(\tilde R_{00})$ and (3.3.25), we obtain

$$\begin{aligned}
&\tilde R_{00}(t)\Theta^*(t)+\tilde B(t)^*P(t)+\tilde S_0(t)=0,\\
&\tilde R_{00}(t)v^*(t)+\tilde B(t)^*\eta(t)+\tilde\rho_0(t)=0,\quad \text{a.e. } t\in[s,T].
\end{aligned}\tag{3.3.27}$$

Step 1: we aim to prove that

$$J(s,\xi;\Theta^*(\cdot)\mathbf{X}^*(\cdot)+v^*(\cdot)) = \big\langle P(s)\xi,\xi\big\rangle + 2\big\langle \eta(s),\xi\big\rangle + \int_s^T \Big\{2\big\langle \eta(t),\tilde{b}(t)\big\rangle - \big\langle \tilde{R}_{00}(t)^\dagger\big[\tilde{B}(t)^*\eta(t)+\tilde{\rho}_0(t)\big], \tilde{B}(t)^*\eta(t)+\tilde{\rho}_0(t)\big\rangle\Big\}dt, \tag{3.3.28}$$

where

$$\mathbf{X}^*(t) = \Phi_\Theta(t,s)\xi + \int_s^t \Phi_\Theta(t,r)\big[\tilde{B}(r)v^*(r)+\tilde{b}(r)\big]dr, \quad t\in[s,T]. \tag{3.3.29}$$

Using (3.3.22)(b), (3.3.29) and by some calculations, we have

$$\begin{aligned}\big\langle P(T)\mathbf{X}^*(T),\mathbf{X}^*(T)\big\rangle = \big\langle P(s)\xi,\xi\big\rangle + \int_s^T \Big\{2\big\langle P(t)\mathbf{X}^*(t), \tilde{B}(t)v^*(t)\\ +\tilde{b}(t)\big\rangle - \langle\big[\tilde{Q}(t)+\Theta^*(t)^*\tilde{S}_0(t)+\tilde{S}_0(t)^*\Theta^*(t)\\ +\Theta^*(t)^*\tilde{R}_{00}(t)\Theta^*(t)\big]\mathbf{X}^*(t),\mathbf{X}^*(t)\big\rangle\Big\}dt.\end{aligned} \tag{3.3.30}$$

Similarly by (3.3.23)(a) and (3.3.29), we get

$$\begin{aligned}2\big\langle \eta(T),\mathbf{X}^*(T)\big\rangle = 2\big\langle \eta(s),\xi\big\rangle + 2\int_s^T \Big[\big\langle \eta(t), \tilde{B}(t)v^*(t)+\tilde{b}(t)\big\rangle\\ -\big\langle \mathbf{X}^*(t), \tilde{q}(t)+P(t)\tilde{b}(t)+\Theta^*(t)^*\tilde{\rho}_0(t)\big\rangle\Big]dt.\end{aligned} \tag{3.3.31}$$

By definition of $J(s,\xi;\Theta^*(\cdot)\mathbf{X}^*(\cdot)+v^*(\cdot))$, (3.3.30) and (3.3.31), we deduce

$$\begin{aligned}&J(s,\xi;\Theta^*(\cdot)\mathbf{X}^*(\cdot)+v^*(\cdot)) = \big\langle P(s)\xi,\xi\big\rangle + 2\big\langle \eta(s),\xi\big\rangle\\ &+\int_s^T \Big\{2\big\langle \eta(t),\tilde{b}(t)\big\rangle + 2\big\langle v^*(t), \big[\tilde{R}_{00}(t)\Theta^*(t)+\tilde{B}(t)^*P(t)\\ &+\tilde{S}_0(t)\big]\mathbf{X}^*(t)+\tilde{\rho}_0(t)+\tilde{B}(t)^*\eta(t)\big\rangle + \big\langle \tilde{R}_{00}(t)v^*(t),v^*(t)\big\rangle\Big\}dt\\ &= \big\langle P(s)\xi,\xi\big\rangle + 2\big\langle \eta(s),\xi\big\rangle + \int_s^T \Big\{2\big\langle \eta(t),\tilde{b}(t)\big\rangle + \Big\langle \tilde{R}_{00}(t)\big[v^*(t)\\ &+\tilde{R}_{00}(t)^\dagger\big[\tilde{B}(t)^*\eta(t)+\tilde{\rho}_0(t)\big]\big], v^*(t)+\tilde{R}_{00}(t)^\dagger\big[\tilde{B}(t)^*\eta(t)+\tilde{\rho}_0(t)\big]\Big\rangle\\ &-\big\langle \tilde{R}_{00}(t)^\dagger\big[\tilde{B}(t)^*\eta(t)+\tilde{\rho}_0(t)\big], \tilde{B}(t)^*\eta(t)+\tilde{\rho}_0(t)\big\rangle\Big\}dt.\end{aligned} \tag{3.3.32}$$

From (3.3.25) we obtain

$$\tilde{R}_{00}(t)\Big\{v^*(t) + \tilde{R}_{00}(t)^\dagger\big[\tilde{B}(t)^*\eta(t) + \tilde{\rho}_0(t)\big]\Big\} = 0, \quad \text{a.e. } t \in [s, T],$$

which and (3.3.32) imply (3.3.28).

By definition for any $\xi \in M^2$ and $u(\cdot) \in L^2(s, T; \mathbb{R}^m)$,

$$\begin{aligned} J(s, \xi; u(\cdot)) = \int_s^T &\Big\{\big\langle \tilde{Q}(t)\mathbf{X}(t), \mathbf{X}(t)\big\rangle + 2\big\langle \tilde{S}_0(t)\mathbf{X}(t), u(t)\big\rangle \\ &+\big\langle \tilde{R}_{00}(t)u(t), u(t)\big\rangle + 2\big\langle \tilde{q}(t), \mathbf{X}(t)\big\rangle + 2\big\langle \tilde{\rho}_0(t), u(t)\big\rangle\Big\}dt \\ &+\big\langle \tilde{G}\mathbf{X}(T), \mathbf{X}(T)\big\rangle + 2\big\langle \tilde{g}, \mathbf{X}(T)\big\rangle, \end{aligned} \tag{3.3.33}$$

where

$$\mathbf{X}(t) = \Phi(t, s)\xi + \int_s^t \Phi(t, r)\big[\tilde{B}(r)u(r) + \tilde{b}(r)\big]dr, \quad t \in [s, T]. \tag{3.3.34}$$

Step 2: we show that

$$\begin{aligned} \big\langle P(T)\mathbf{X}(T), \mathbf{X}(T)\big\rangle = \big\langle P(s)\xi, \xi\big\rangle + \int_s^T &\Big\{2\big\langle P(t)\mathbf{X}(t), \tilde{B}(t)u(t) \\ &+\tilde{b}(t) - \tilde{B}(t)\Theta^*(t)\mathbf{X}(t)\big\rangle - \big\langle\big[\tilde{Q}(t) + \Theta^*(t)^*\tilde{S}_0(t) \\ &+\tilde{S}_0(t)^*\Theta^*(t) + \Theta^*(t)^*\tilde{R}_{00}(t)\Theta^*(t)\big]\mathbf{X}(t), \mathbf{X}(t)\big\rangle\Big\}dt, \end{aligned} \tag{3.3.35}$$

and

$$\begin{aligned} 2\big\langle \eta(T), \mathbf{X}(T)\big\rangle = 2\big\langle \eta(s), \xi\big\rangle + 2\int_s^T &\Big\{\big\langle \eta(t), \tilde{B}(t)u(t) + \tilde{b}(t) \\ &-\tilde{B}(t)\Theta^*(t)\mathbf{X}(t)\big\rangle - \big\langle \mathbf{X}(t), \tilde{q}(t) + P(t)\tilde{b}(t) + \Theta^*(t)^*\tilde{\rho}_0(t)\big\rangle\Big\}dt. \end{aligned} \tag{3.3.36}$$

By (3.3.22)(a) and (3.3.25), we have

$$\begin{aligned} \big\langle P(s)\xi, \xi\big\rangle &= \big\langle \tilde{G}\Phi(T, s)\xi, \Phi(T, s)\xi\big\rangle \\ &+\int_s^T \big\langle \Phi(r, s)\xi, \big[\tilde{Q}(r) - \Theta^*(r)^*\tilde{R}_{00}(r)\Theta^*(r)\big]\Phi(r, s)\xi\big\rangle dr, \end{aligned} \tag{3.3.37}$$

and

$$
\begin{aligned}
&2\int_s^T \big\langle P(t)\mathbf{X}(t),\, \tilde{B}(t)u(t)+\tilde{b}(t)-\tilde{B}(t)\Theta^*(t)\mathbf{X}(t)\big\rangle dt\\
&=2\int_s^T \Big\langle \Phi(T,t)\big[\tilde{B}(t)u(t)+\tilde{b}(t)\big],\, \tilde{G}\Phi(T,s)\xi\\
&\qquad +\tilde{G}\int_s^t \Phi(T,r)\big[\tilde{B}(r)u(r)+\tilde{b}(r)\big]dr\Big\rangle dt\\
&+2\int_s^T\int_t^T \Big\langle \Phi(r,t)\big[\tilde{B}(t)u(t)+\tilde{b}(t)\big],\big[\tilde{Q}(r)-\Theta^*(r)^*\tilde{R}_{00}(r)\Theta^*(r)\big]\Phi(r,s)\xi\\
&\qquad +\big[\tilde{Q}(r)-\Theta^*(r)^*\tilde{R}_{00}(r)\Theta^*(r)\big]\int_s^t \Phi(r,\alpha)\big[\tilde{B}(\alpha)u(\alpha)+\tilde{b}(\alpha)\big]d\alpha\Big\rangle drdt\\
&-2\int_s^T \big\langle P(t)\mathbf{X}(t),\, \tilde{B}(t)\Theta^*(t)\mathbf{X}(t)\big\rangle dt.
\end{aligned}
\tag{3.3.38}
$$

Recalling (3.3.34), we have

$$
\begin{aligned}
&-\int_s^T \langle[\tilde{Q}(t)+\Theta^*(t)^*\tilde{S}_0(t)+\tilde{S}_0(t)^*\Theta^*(t)\\
&\qquad +\Theta^*(t)^*\tilde{R}_{00}(t)\Theta^*(t)\big]\mathbf{X}(t),\mathbf{X}(t)\rangle dt\\
&=-\int_s^T \langle[\tilde{Q}(t)+\Theta^*(t)^*\tilde{S}_0(t)+\tilde{S}_0(t)^*\Theta^*(t)\\
&\qquad +\Theta^*(t)^*\tilde{R}_{00}(t)\Theta^*(t)\big]\Phi(t,s)\xi,\Phi(t,s)\xi\rangle dt\\
&\quad -2\int_s^T \Big\langle\big[\tilde{Q}(t)+\Theta^*(t)^*\tilde{S}_0(t)+\tilde{S}_0(t)^*\Theta^*(t)+\Theta^*(t)^*\tilde{R}_{00}(t)\Theta^*(t)\big]\\
&\qquad \times\Phi(t,s)\xi,\int_s^t \Phi(t,r)\big[\tilde{B}(r)u(r)+\tilde{b}(r)\big]dr\Big\rangle dt\\
&-\int_s^T \Big\langle\big[\tilde{Q}(t)+\Theta^*(t)^*\tilde{S}_0(t)+\tilde{S}_0(t)^*\Theta^*(t)+\Theta^*(t)^*\tilde{R}_{00}(t)\Theta^*(t)\big]\\
&\qquad \times\int_s^t \Phi(t,r)\big[\tilde{B}(r)u(r)+\tilde{b}(r)\big]dr,\int_s^t \Phi(t,r)\big[\tilde{B}(r)u(r)+\tilde{b}(r)\big]dr\Big\rangle dt.
\end{aligned}
\tag{3.3.39}
$$

Noting (3.3.27), we deduce

$$
\begin{aligned}
&\int_s^T \langle \Phi(t,s)\xi, [\tilde{Q}(t) - \Theta^*(t)^*\tilde{R}_{00}(t)\Theta^*(t)]\Phi(t,s)\xi\rangle dt \\
&-\int_s^T \langle [\tilde{Q}(t) + \Theta^*(t)^*\tilde{S}_0(t) + \tilde{S}_0(t)^*\Theta^*(t) \\
&\quad +\Theta^*(t)^*\tilde{R}_{00}(t)\Theta^*(t)]\Phi(t,s)\xi, \Phi(t,s)\xi\rangle dt \\
=& -\int_s^T \langle [\Theta^*(t)^*\tilde{R}_{00}(t)\Theta^*(t) + \Theta^*(t)^*\tilde{S}_0(t) + \tilde{S}_0(t)^*\Theta^*(t) \\
&\quad +\Theta^*(t)^*\tilde{R}_{00}(t)\Theta^*(t)]\Phi(t,s)\xi, \Phi(t,s)\xi\rangle dt \\
=& \int_s^T \langle [\Theta^*(t)^*\tilde{B}(t)^*P(t) + P(t)\tilde{B}(t)\Theta^*(t)]\Phi(t,s)\xi, \Phi(t,s)\xi\rangle dt,
\end{aligned}
\tag{3.3.40}
$$

$$
\begin{aligned}
&2\int_s^T \int_t^T \langle \Phi(r,t)[\tilde{B}(t)u(t) + \tilde{b}(t)], [\tilde{Q}(r) \\
&\quad -\Theta^*(r)^*\tilde{R}_{00}(r)\Theta^*(r)]\Phi(r,s)\xi\rangle dr dt \\
&-2\int_s^T \Big\langle [\tilde{Q}(t) + \Theta^*(t)^*\tilde{S}_0(t) + \tilde{S}_0(t)^*\Theta^*(t) + \Theta^*(t)^*\tilde{R}_{00}(t)\Theta^*(t)] \\
&\quad \times\Phi(t,s)\xi, \int_s^t \Phi(t,r)[\tilde{B}(r)u(r) + \tilde{b}(r)]dr\Big\rangle dt \\
=& 2\int_s^T \int_t^T \langle \Phi(r,t)[\tilde{B}(t)u(t) + \tilde{b}(t)], [\Theta^*(r)^*\tilde{B}(r)^*P(r) \\
&\quad +P(r)\tilde{B}(r)\Theta^*(r)]\Phi(r,s)\xi\rangle dr dt,
\end{aligned}
\tag{3.3.41}
$$

and

$$
\begin{aligned}
&2\int_s^T \int_t^T \Big\langle \Phi(r,t)[\tilde{B}(t)u(t) + \tilde{b}(t)], [\tilde{Q}(r) - \Theta^*(r)^*\tilde{R}_{00}(r)\Theta^*(r)] \\
&\quad \times\int_s^t \Phi(r,\alpha)[\tilde{B}(\alpha)u(\alpha) + \tilde{b}(\alpha)]d\alpha\Big\rangle dr dt \\
&-\int_s^T \Big\langle [\tilde{Q}(t) + \Theta^*(t)^*\tilde{S}_0(t) + \tilde{S}_0(t)^*\Theta^*(t) + \Theta^*(t)^*\tilde{R}_{00}(t)\Theta^*(t)] \\
&\quad \times\int_s^t \Phi(t,r)[\tilde{B}(r)u(r) + \tilde{b}(r)]dr, \int_s^t \Phi(t,r)[\tilde{B}(r)u(r) + \tilde{b}(r)]dr\Big\rangle dt \\
=& \int_s^T \Big\langle [\Theta^*(t)^*\tilde{B}(t)^*P(t) + P(t)\tilde{B}(t)\Theta^*(t)] \int_s^t \Phi(t,r)[\tilde{B}(r)u(r) \\
&\quad +\tilde{b}(r)]dr, \int_s^t \Phi(t,r)[\tilde{B}(r)u(r) + \tilde{b}(r)]dr\Big\rangle dt.
\end{aligned}
\tag{3.3.42}
$$

By (3.3.34), we have

$$
\begin{aligned}
&\big\langle \tilde{G}\mathbf{X}(T), \mathbf{X}(T)\big\rangle \\
&= \big\langle \tilde{G}\Phi(T,s)\xi, \Phi(T,s)\xi\big\rangle + 2\Big\langle \tilde{G}\Phi(T,s)\xi, \int_s^T \Phi(T,r)\big[\tilde{B}(r)u(r)+\tilde{b}(r)\big]dr\Big\rangle \\
&\quad + \Big\langle \tilde{G}\int_s^T \Phi(T,r)\big[\tilde{B}(r)u(r)+\tilde{b}(r)\big]dr, \int_s^T \Phi(T,r)\big[\tilde{B}(r)u(r)+\tilde{b}(r)\big]dr\Big\rangle,
\end{aligned}
$$

which and (3.3.37)–(3.3.42) imply that (3.3.35) holds. Similarly (3.3.36) can be proved.

Step 3: we prove that $(\Theta^*(\cdot), v^*(\cdot))$ is the optimal closed-loop strategy of Problem (P-NCD) and (3.3.26) is the value function.

Using (3.3.33), (3.3.35) and (3.3.36), we derive

$$
\begin{aligned}
J(s,\xi;u(\cdot)) &= \big\langle P(s)\xi, \xi\big\rangle + 2\big\langle \eta(s), \xi\big\rangle + \int_s^T \Big\{\big\langle \tilde{R}_{00}(t)u(t), u(t)\big\rangle \\
&\quad +2\big\langle u(t), \big[\tilde{B}(t)^*P(t)+\tilde{S}_0(t)\big]\mathbf{X}(t) + \tilde{\rho}_0(t) + \tilde{B}(t)^*\eta(t)\big\rangle \\
&\quad -\big\langle \mathbf{X}(t), \big[2P(t)\tilde{B}(t)\Theta^*(t) + \Theta^*(t)^*\tilde{S}_0(t) + \tilde{S}_0(t)^*\Theta^*(t) \\
&\quad +\Theta^*(t)^*\tilde{R}_{00}(t)\Theta^*(t)\big]\mathbf{X}(t) + 2\Theta^*(t)^*[\tilde{B}(t)^*\eta(t)+\tilde{\rho}_0(t)]\big\rangle \\
&\quad +2\big\langle \eta(t), \tilde{b}(t)\big\rangle\Big\}dt.
\end{aligned}
\tag{3.3.43}
$$

Notice that

$$
\begin{aligned}
&\big\langle\big[2P(t)\tilde{B}(t)\Theta^*(t) + \Theta^*(t)^*\tilde{S}_0(t) + \tilde{S}_0(t)^*\Theta^*(t) \\
&\quad +\Theta^*(t)^*\tilde{R}_{00}(t)\Theta^*(t)\big]\mathbf{X}(\cdot), \mathbf{X}(\cdot)\big\rangle \\
&= 2\big\langle\big[P(t)\tilde{B}(t)+\tilde{S}_0(t)^*\big]\Theta^*(t)\mathbf{X}(t), \mathbf{X}(t)\big\rangle + \big\langle \tilde{R}_{00}(t)\Theta^*(t)\mathbf{X}(t), \Theta^*(t)\mathbf{X}(t)\big\rangle \\
&= -\big\langle \tilde{R}_{00}(t)\Theta^*(t)\mathbf{X}(t), \Theta^*(t)\mathbf{X}(t)\big\rangle, \quad \text{a.e. } t\in[s,T],
\end{aligned}
$$

which and (3.3.43) imply that

$$
\begin{aligned}
J(s,\xi;u(\cdot)) &= J(s,\xi;\Theta^*(\cdot)\mathbf{X}^*(\cdot)+v^*(\cdot)) \\
&\quad + \int_s^T \Big\{\big\langle \tilde{R}_{00}(t)^\dagger\tilde{R}_{00}(t)v^*(t), \tilde{R}_{00}(t)v^*(t)\big\rangle \\
&\quad + \big\langle \tilde{R}_{00}(t)u(t), u(t)\big\rangle - 2\big\langle u(t), \tilde{R}_{00}(t)\big[\Theta^*(t)\mathbf{X}(t)+v^*(t)\big]\big\rangle \\
&\quad + \big\langle \tilde{R}_{00}(t)\Theta^*(t)\mathbf{X}(t), \Theta^*(t)\mathbf{X}(t)\big\rangle + 2\big\langle \tilde{R}_{00}(t)\Theta^*(t)\mathbf{X}(t), v^*(t)\big\rangle\Big\}dt \\
&= J(s,\xi;\Theta^*(\cdot)\mathbf{X}^*(\cdot)+v^*(\cdot)) + \int_s^T \big\langle \tilde{R}_{00}(t)\big[u(t)-\Theta^*(t)\mathbf{X}(t)
\end{aligned}
$$

$$- v^*(t)\big], u(t) - \Theta^*(t)\mathbf{X}(t) - v^*(t)\big\rangle dt$$
$$\geqslant J(s,\xi;\Theta^*(\cdot)\mathbf{X}^*(\cdot) + v^*(\cdot)). \tag{3.3.44}$$

Since (3.3.44) holds for any $\xi \in M^2$, $u(\cdot) \in L^2(s,T;\mathbb{R}^m)$, $(\Theta^*(\cdot), v^*(\cdot))$ is the optimal closed-loop strategy of Problem (P-NCD) and (3.3.26) is proved. This completes the proof of Theorem 3.3.7. □

The following lemma reveals the relationship among the equations involved in the above theorem.

Lemma 3.3.8 *Let* (H3.3) *hold. Suppose* $\tilde{S}_0^*[I-\tilde{R}_{00}^\dagger\tilde{R}_{00}] = 0$, $R_{00}^\dagger \in L^\infty(s,T;\mathbb{R}^m)$, *then the following three conditions are equivalent:*

(i) *(3.3.22)(a) admits a unique solution $P(\cdot)$ in the class of strongly continuous self adjoint operators in M^2, $\mathcal{R}(\tilde{B}^*P + \tilde{S}_0) \subseteq \mathcal{R}(\tilde{R}_{00})$;*
(ii) *(3.3.22)(b) admits a unique solution $P(\cdot)$ in the class of strongly continuous self adjoint operators in M^2, where $\Theta^*(\cdot)$ is given by (3.3.25)(a) and $\Phi_\Theta(\cdot,\cdot)$ satisfies (3.3.24), $\mathcal{R}(\tilde{B}^*P + \tilde{S}_0) \subseteq \mathcal{R}(\tilde{R}_{00})$;*
(iii) *(3.3.22)(c) admits a unique solution $P(\cdot)$ in the class of strongly continuous self adjoint operators in M^2, where $\Theta^*(\cdot)$ is given by (3.3.25)(a) and $\Phi_\Theta(\cdot,\cdot)$ satisfies (3.3.24), $\mathcal{R}(\tilde{B}^*P + \tilde{S}_0) \subseteq \mathcal{R}(\tilde{R}_{00})$;*

In addition, (3.3.23)(a) and (3.3.23)(b) are also equivalent.

Proof In the following proof, we simply write $\langle\cdot,\cdot\rangle_{M^2}$, $\langle\cdot,\cdot\rangle_{L^2}$ and $\langle\cdot,\cdot\rangle_Z$ as $\langle\cdot,\cdot\rangle$ if no ambiguity exists. To prove that $(i) \Leftrightarrow (ii) \Leftrightarrow (iii)$, we only need to prove that $(i) \Leftrightarrow (iii)$ and $(ii) \Leftrightarrow (iii)$.

(a) $(i) \Rightarrow (iii)$. Suppose that (3.3.22)(a) admits a unique solution $P(\cdot)$ in the class of strongly continuous self adjoint operators in M^2, and $\tilde{P}(\cdot)$ satisfies the following integral equation:

$$\begin{aligned}\big\langle \tilde{P}(t)\xi, \xi'\big\rangle &= \big\langle \Phi(T,t)\xi', \tilde{G}\Phi_\Theta(T,t)\xi\big\rangle \\ &\quad + \int_t^T \Big\langle \Phi(r,t)\xi', \Big\{\tilde{Q}(r) - \tilde{S}_0(r)^*\tilde{R}_{00}(r)^\dagger\big[\tilde{B}(r)^*P(r) \\ &\qquad + \tilde{S}_0(r)\big]\Big\}\Phi_\Theta(r,t)\xi\Big\rangle dr, \quad t \in [s,T], \quad \xi,\xi' \in M^2,\end{aligned} \tag{3.3.45}$$

where $\Theta^*(\cdot)$ is given by (3.3.25)(a) and $\Phi_\Theta(\cdot,\cdot)$ satisfies (3.3.24). By (3.3.25)(a) and $\mathcal{R}(\tilde{B}^*P + \tilde{S}_0) \subseteq \mathcal{R}(\tilde{R}_{00})$, we have

$$\tilde{R}_{00}(r)\Theta^*(r) + \tilde{B}(r)^*P(r) + \tilde{S}_0(r) = 0, \quad \text{a.e. } r \in [s,T]. \tag{3.3.46}$$

Noting $\tilde{R}_{00}^{\dagger} = R_{00}^{\dagger} \in \mathcal{L}(\mathbb{R}^m)$, $(\tilde{R}_{00}^{\dagger})^* = \tilde{R}_{00}^{\dagger}$ and $\tilde{S}_0^*[I - \tilde{R}_{00}^{\dagger}\tilde{R}_{00}] = 0$, we deduce

$$
\begin{aligned}
&\tilde{Q}(r) - \big[\tilde{R}_{00}(r)^{\dagger}\big(\tilde{B}(r)^* P(r) + \tilde{S}_0(r)\big)\big]^*\big[\tilde{B}(r)^* P(r) + \tilde{S}_0(r)\big]\\
&= \tilde{Q}(r) - \tilde{S}_0(r)^* \tilde{R}_{00}(r)^{\dagger}\big[\tilde{B}(r)^* P(r) + \tilde{S}_0(r)\big]\\
&\quad + \tilde{S}_0(r)^* \tilde{R}_{00}(r)^{\dagger}\big[\tilde{B}(r)^* P(r) + \tilde{S}_0(r)\big]\\
&\quad - \big[\tilde{R}_{00}(r)^{\dagger}\big(\tilde{B}(r)^* P(r) + \tilde{S}_0(r)\big)\big]^*\big[\tilde{B}(r)^* P(r) + \tilde{S}_0(r)\big]\\
&= \tilde{Q}(r) - \tilde{S}_0(r)^* \tilde{R}_{00}(r)^{\dagger}\big[\tilde{B}(r)^* P(r) + \tilde{S}_0(r)\big]\\
&\quad - \big[\tilde{B}(r)^* P(r)\big]^* \tilde{R}_{00}(r)^{\dagger}\big[\tilde{B}(r)^* P(r) + \tilde{S}_0(r)\big]\\
&= \tilde{Q}(r) - \tilde{S}_0(r)^* \tilde{R}_{00}(r)^{\dagger}\big[\tilde{B}(r)^* P(r) + \tilde{S}_0(r)\big]\\
&\quad + \big[\tilde{B}(r)^* P(r)\big]^*\big[\Theta^*(r) - \big(I - \tilde{R}_{00}(r)^{\dagger}\tilde{R}_{00}(r)\big)\theta(r)\big]\\
&\quad - \tilde{S}_0(r)^*\big(I - \tilde{R}_{00}(r)^{\dagger}\tilde{R}_{00}(r)\big)\theta(r)\\
&= \tilde{Q}(r) - \tilde{S}_0(r)^* \tilde{R}_{00}(r)^{\dagger}\big[\tilde{B}(r)^* P(r) + \tilde{S}_0(r)\big]\\
&\quad + \big[\tilde{B}(r)^* P(r)\big]^*\Theta^*(r) + \Theta^*(r)^*\tilde{R}_{00}(r)\big[I - \tilde{R}_{00}(r)^{\dagger}\tilde{R}_{00}(r)\big]\theta(r)\\
&= \tilde{Q}(r) - \tilde{S}_0(r)^* \tilde{R}_{00}(r)^{\dagger}\big[\tilde{B}(r)^* P(r) + \tilde{S}_0(r)\big]\\
&\quad + (\tilde{B}(r)^* P(r))^*\Theta^*(r), \qquad \text{a.e. } r \in [s, T].
\end{aligned}
\tag{3.3.47}
$$

Noting

$$
\begin{aligned}
\big\langle P(t)\xi, \xi'\big\rangle = &\big\langle \Phi(T,t)\xi', \tilde{G}\Phi(T,t)\xi\big\rangle\\
&+ \int_t^T \Big\langle \Phi(r,t)\xi', \Big\{\tilde{Q}(r) - \Big(\tilde{R}_{00}(r)^{\dagger}\big[\tilde{B}(r)^* P(r)\\
&+ \tilde{S}_0(r)\big]\Big)^*\big[\tilde{B}(r)^* P(r) + \tilde{S}_0(r)\big]\Big\}\Phi(r,t)\xi\Big\rangle dr,
\end{aligned}
\tag{3.3.48}
$$

by (3.3.45)–(3.3.48) we derive

$$
\begin{aligned}
&\big\langle[\tilde{P}(t) - P(t)]\xi, \xi'\big\rangle = \Big\langle \Phi(T,t)\xi', \tilde{G}\int_t^T \Phi_{\Theta}(T,r)\tilde{B}(r)\Theta^*(r)\Phi(r,t)\xi dr\Big\rangle\\
&+ \int_t^T \big\langle \Phi(r,t)\xi', -\big[\tilde{B}(r)^* P(r)\big]^*\Theta^*(r)\Phi(r,t)\xi\big\rangle dr\\
&+ \int_t^T \Big\langle \Phi(r,t)\xi', \Big\{\tilde{Q}(r) - \Big(\tilde{R}_{00}(r)^{\dagger}\big[\tilde{B}(r)^* P(r) + \tilde{S}_0(r)\big]\Big)^*\big[\tilde{B}(r)^* P(r)\\
&+ \tilde{S}_0(r)\big] - \big[\tilde{B}(r)^* P(r)\big]^*\Theta^*(r)\Big\}\int_t^r \Phi_{\Theta}(r,\alpha)\tilde{B}(\alpha)\Theta^*(\alpha)\Phi(\alpha,t)\xi d\alpha\Big\rangle dr.
\end{aligned}
\tag{3.3.49}
$$

Noting

$$
\begin{aligned}
&\int_t^T \Big\langle \Phi(r,t)\xi', \Big\{\tilde{Q}(r) - \Big(\tilde{R}_{00}(r)^{\dagger}\big[\tilde{B}(r)^*P(r) + \tilde{S}_0(r)\big]\Big)^*\big[\tilde{B}(r)^*P(r) \\
&\quad +\tilde{S}_0(r)\big] - \big[\tilde{B}(r)^*P(r)\big]^*\Theta^*(r)\Big\} \int_t^r \Phi_\Theta(r,\alpha)\tilde{B}(\alpha)\Theta^*(\alpha)\Phi(\alpha,t)\xi d\alpha\Big\rangle dr \\
&= \int_t^T \int_r^T \Big\langle \Phi_\Theta(\alpha,r)^*\Big\{\tilde{Q}(\alpha) - \big[\tilde{B}(\alpha)^*P(\alpha) \\
&\quad +\tilde{S}_0(\alpha)\big]^*\tilde{R}_{00}(\alpha)^{\dagger}\big[\tilde{B}(\alpha)^*P(\alpha) \\
&\quad +\tilde{S}_0(\alpha)\big] - \Theta^*(\alpha)^*\tilde{B}(\alpha)^*P(\alpha)\Big\}\Phi(\alpha,t)\xi', \tilde{B}(r)\Theta^*(r)\Phi(r,t)\xi\Big\rangle d\alpha dr,
\end{aligned}
\tag{3.3.50}
$$

and

$$
\begin{aligned}
&\int_t^T \Big\langle \tilde{B}(r)\Theta^*(r)\Phi(r,t)\xi, \Phi_\Theta(T,r)^*\tilde{G}\Phi(T,t)\xi' \\
&\quad + \int_r^T \Phi_\Theta(\alpha,r)^*\Big\{\tilde{Q}(\alpha) - \big[\tilde{B}(\alpha)^*P(\alpha) + \tilde{S}_0(\alpha)\big]^*\tilde{R}_{00}(\alpha)^{\dagger}\big[\tilde{B}(\alpha)^*P(\alpha) \\
&\qquad +\tilde{S}_0(\alpha)\big] - \Theta^*(\alpha)^*\tilde{B}(\alpha)^*P(\alpha)\Big\}\Phi(\alpha,t)\xi' d\alpha\Big\rangle dr \\
&= \int_t^T \langle \tilde{P}(r)\tilde{B}(r)\Theta^*(r)\Phi(r,t)\xi, \Phi(r,t)\xi'\rangle dr,
\end{aligned}
$$

which and (3.3.49), (3.3.50) imply that for any $\xi, \xi' \in M^2$,

$$
\big\langle[\tilde{P}(t) - P(t)]\xi, \xi'\big\rangle = \int_t^T \langle \Phi(r,t)^*\big[\tilde{P}(r) - P(r)\big]\tilde{B}(r)\Theta^*(r)\Phi(r,t)\xi, \xi'\rangle dr.
$$

Thus $P(t) = \tilde{P}(t)$ for all $t \in [s,T]$.

(b) $(iii) \Rightarrow (i)$. Suppose that (3.3.22)(c) admits a unique solution $P(\cdot)$ in the class of strongly continuous self adjoint operators in M^2, where $\Theta^*(\cdot)$ is given by (3.3.25)(a) and $\Phi_\Theta(\cdot,\cdot)$ satisfies (3.3.24), and $\tilde{P}(\cdot)$ satisfies the following integral equation:

$$
\begin{aligned}
\big\langle \tilde{P}(t)\xi, \xi'\big\rangle &= \big\langle \Phi(T,t)\xi', \tilde{G}\Phi(T,t)\xi\big\rangle \\
&\quad + \int_t^T \langle \Phi(r,t)\xi', \big[\tilde{Q}(r) - \Theta^*(r)^*\tilde{R}_{00}(r)\Theta^*(r)\big]\Phi(r,t)\xi\rangle dr.
\end{aligned}
\tag{3.3.51}
$$

By (3.3.25)(a) and $\mathcal{R}(\tilde{B}^*P + \tilde{S}_0) \subseteq \mathcal{R}(\tilde{R}_{00})$, we have

$$
\tilde{R}_{00}(r)\Theta^*(r) + \tilde{B}(r)^*P(r) + \tilde{S}_0(r) = 0, \quad \text{a.e. } r \in [s,T].
$$

Noting $\tilde{S}_0^*\big[I - \tilde{R}_{00}^{\dagger}\tilde{R}_{00}\big] = 0$, we deduce

$$\begin{aligned}
&\tilde{Q}(r) + \Theta^*(r)^*\tilde{R}_{00}(r)\tilde{R}_{00}(r)^{\dagger}\tilde{S}_0(r)\\
&= \tilde{Q}(r) - \Theta^*(r)^*\tilde{R}_{00}(r)\Theta^*(r) + \Theta^*(r)^*\big[\tilde{R}_{00}(r)\Theta^*(r)\\
&\quad +\tilde{R}_{00}(r)\tilde{R}_{00}(r)^{\dagger}\tilde{S}_0(r)\big]\\
&= \tilde{Q}(r) - \Theta^*(r)^*\tilde{R}_{00}(r)\Theta^*(r) - \Theta^*(r)^*\tilde{B}(r)^*P(r).
\end{aligned} \tag{3.3.52}$$

Noting

$$\begin{aligned}
&\big\langle P(t)\xi, \xi'\big\rangle = \big\langle \Phi(T,t)\xi', \tilde{G}\Phi_\Theta(T,t)\xi\big\rangle\\
&\quad + \int_t^T \Big\langle \Phi(r,t)\xi', \Big\{\tilde{Q}(r) - \tilde{S}_0(r)^*\tilde{R}_{00}(r)^{\dagger}\big[\tilde{B}(r)^*P(r)\\
&\quad +\tilde{S}_0(r)\big]\Big\}\Phi_\Theta(r,t)\xi\Big\rangle dr\\
&= \Big\langle \tilde{G}\Phi(T,t)\xi', \Phi(T,t)\xi + \int_t^T \Phi(T,r)\tilde{B}(r)\Theta^*(r)\Phi_\Theta(r,t)\xi dr\Big\rangle\\
&\quad + \int_t^T \Big\langle \big[\tilde{Q}(r) + \Theta^*(r)^*\tilde{R}_{00}(r)\tilde{R}_{00}(r)^{\dagger}\tilde{S}_0(r)\big]\Phi(r,t)\xi', \Phi(r,t)\xi\\
&\quad + \int_t^r \Phi(r,\alpha)\tilde{B}(\alpha)\Theta^*(\alpha)\Phi_\Theta(\alpha,t)\xi d\alpha\Big\rangle dr,\ t\in[s,T],\ \xi,\xi'\in M^2,
\end{aligned} \tag{3.3.53}$$

by (3.3.51), (3.3.52) and (3.3.53) we derive

$$\begin{aligned}
&\langle [P(t) - \tilde{P}(t)]\xi, \xi'\rangle\\
&= -\int_t^T \big\langle P(r)\Phi(r,t)\xi', \tilde{B}(r)\Theta^*(r)\Phi(r,t)\xi\big\rangle dr\\
&\quad + \int_t^T \Big\langle \tilde{B}(r)\Theta^*(r)\Phi_\Theta(r,t)\xi, \Phi(T,r)^*\tilde{G}\Phi(T,t)\xi'\\
&\quad + \int_r^T \Phi(\alpha,r)^*\big[\tilde{Q}(\alpha) + \Theta^*(\alpha)^*\tilde{R}_{00}(\alpha)\tilde{R}_{00}(\alpha)^{\dagger}\tilde{S}_0(\alpha)\big]\Phi(\alpha,t)\xi' d\alpha\Big\rangle dr\\
&= -\int_t^T \langle [P(r) - \tilde{P}(r)]\Phi(r,t)\xi', \tilde{B}(r)\Theta^*(r)\Phi(r,t)\xi\big\rangle dr\\
&\quad - \int_t^T \big\langle \tilde{G}\Phi(T,t)\xi', \Phi(T,r)\tilde{B}(r)\Theta^*(r)\big[\Phi(r,t) - \Phi_\Theta(r,t)\big]\xi\big\rangle dr
\end{aligned}$$

$$
\begin{aligned}
&-\int_t^T\int_r^T \langle [\tilde{Q}(\alpha) - \Theta^*(\alpha)^*\tilde{R}_{00}(\alpha)\Theta^*(\alpha)]\Phi(\alpha,t)\xi', \\
&\qquad \Phi(\alpha,r)\tilde{B}(r)\Theta^*(r)[\Phi(r,t) - \Phi_\Theta(r,t)]\xi\rangle d\alpha dr \\
&+\int_t^T\int_r^T \langle \tilde{B}(r)\Theta^*(r)\Phi_\Theta(r,t)\xi, \Phi(\alpha,r)^*[\tilde{Q}(\alpha) + \Theta^*(\alpha)^*\tilde{R}_{00}(\alpha)\tilde{R}_{00}(\alpha)^\dagger \\
&\qquad \times \tilde{S}_0(\alpha) - \tilde{Q}(\alpha) + \Theta^*(\alpha)^*\tilde{R}_{00}(\alpha)\Theta^*(\alpha)]\Phi(\alpha,t)\xi'\rangle d\alpha dr \\
=&-\int_t^T \langle [P(r) - \tilde{P}(r)]\Phi(r,t)\xi', \tilde{B}(r)\Theta^*(r)\Phi(r,t)\xi\rangle dr \\
&+\int_t^T \langle \tilde{P}(r)\Phi(r,t)\xi', \tilde{B}(r)\Theta^*(r)[\Phi_\Theta(r,t) - \Phi(r,t)]\xi\rangle dr \\
&-\int_t^T\int_r^T \langle \tilde{B}(r)\Theta^*(r)\Phi_\Theta(r,t)\xi, \Phi(\alpha,r)^*\Theta^*(\alpha)^*\tilde{B} \\
&\times(\alpha)^*P(\alpha)\Phi(\alpha,t)\xi'\rangle d\alpha dr,
\end{aligned}
$$

which yields

$$
\begin{aligned}
&\langle [P(t) - \tilde{P}(t)]\xi, \xi'\rangle \\
&= -\int_t^T \langle [P(r) - \tilde{P}(r)]\Phi(r,t)\xi', \tilde{B}(r)\Theta^*(r)\Phi(r,t)\xi\rangle dr \\
&\quad -\int_t^T \langle [P(r) - \tilde{P}(r)]\Phi(r,t)\xi', \tilde{B}(r)\Theta^*(r)[\Phi_\Theta(r,t) - \Phi(r,t)]\xi\rangle dr \\
&= -\int_t^T \langle [P(r) - \tilde{P}(r)]\Phi(r,t)\xi', \tilde{B}(r)\Theta^*(r)\Phi_\Theta(r,t)\xi\rangle dr, \\
&\quad \text{for any } \xi, \xi' \in M^2,
\end{aligned}
$$

thus $P(t) = \tilde{P}(t)$ for all $t \in [s, T]$.

(c) $(ii) \Rightarrow (iii)$. Supposse (3.3.22)(b) admits a unique solution $P(\cdot)$ in the class of strongly continuous self adjoint operators in M^2, where $\Theta^*(\cdot)$ is given by (3.3.25)(a) and $\Phi_\Theta(\cdot,\cdot)$ satisfies (3.3.24), then we have

$$
\begin{aligned}
\langle P(t)\xi, \xi'\rangle = &\langle \tilde{G}\Phi_\Theta(T,t)\xi, \Phi_\Theta(T,t)\xi'\rangle + \int_t^T \langle [\tilde{Q}(r) + \Theta^*(r)^*\tilde{S}_0(r) \\
&+\tilde{S}_0(r)^*\Theta^*(r) + \Theta^*(r)^*\tilde{R}_{00}(r)\Theta^*(r)]\Phi_\Theta(r,t)\xi, \Phi_\Theta(r,t)\xi'\rangle dr,
\end{aligned}
$$

by (3.3.24)(b) and $\tilde{S}_0^*[I - \tilde{R}_{00}^\dagger \tilde{R}_{00}] = 0$, we deduce

$$
\begin{aligned}
&\langle P(t)\xi, \xi' \rangle \\
&= \langle \Phi(T,t)\xi', \tilde{G}\Phi_\Theta(T,t)\xi \rangle + \int_t^T \Big\langle \Phi(r,t)\xi', \Big\{ \tilde{Q}(r) - \tilde{S}_0(r)^* \\
&\quad \times \tilde{R}_{00}(r)^\dagger \big[\tilde{B}(r)^* P(r) + \tilde{S}_0(r)\big]\Big\} \Phi_\Theta(r,t)\xi \Big\rangle dr, \quad \forall\, \xi, \xi' \in M^2,
\end{aligned}
\tag{3.3.54}
$$

which implies that $P(t)$ satisfies (3.3.22)(c).

(d) $(iii) \Rightarrow (ii)$. Suppose that (3.3.22)(c) admits a unique solution $P(\cdot)$ in the class of strongly continuous self adjoint operators in M^2, where $\Theta^*(\cdot)$ is given by (3.3.25)(a) and $\Phi_\Theta(\cdot,\cdot)$ satisfies (3.3.24), and $\tilde{P}(\cdot)$ satisfies the following integral equation:

$$
\begin{aligned}
&\langle \tilde{P}(t)\xi, \xi' \rangle \\
&= \langle \tilde{G}\Phi_\Theta(T,t)\xi, \Phi_\Theta(T,t)\xi' \rangle + \int_t^T \langle [\tilde{Q}(r) + \Theta^*(r)^* \tilde{S}_0(r) \\
&\quad + \tilde{S}_0(r)^* \Theta^*(r) + \Theta^*(r)^* \tilde{R}_{00}(r) \Theta^*(r)] \Phi_\Theta(r,t)\xi, \Phi_\Theta(r,t)\xi' \rangle dr.
\end{aligned}
$$

Noting (3.3.54), it follows that

$$
\begin{aligned}
&\langle [\tilde{P}(t) - P(t)]\xi, \xi' \rangle \\
&= \Big\langle \tilde{G}\Phi_\Theta(T,t)\xi, \int_t^T \Phi_\Theta(T,r)\tilde{B}(r)\Theta^*(r)\Phi(r,t)\xi' dr \Big\rangle \\
&\quad + \int_t^T \Big\langle [\tilde{Q}(r) + \Theta^*(r)^*\tilde{S}_0(r) + \tilde{S}_0(r)^*\Theta^*(r) + \Theta^*(r)^*\tilde{R}_{00}(r)\Theta^*(r)] \\
&\qquad \times \Phi_\Theta(r,t)\xi, \int_t^r \Phi_\Theta(r,\alpha)\tilde{B}(\alpha)\Theta^*(\alpha)\Phi(\alpha,t)\xi' d\alpha \Big\rangle dr \\
&\quad + \int_t^T \Big\langle \Phi_\Theta(r,t)\xi, \Big\{ \tilde{Q}(r) + \Theta^*(r)^*\tilde{S}_0(r) + \tilde{S}_0(r)^*\Theta^*(r) \\
&\qquad + \Theta^*(r)^*\tilde{R}_{00}(r)\Theta^*(r) - \tilde{Q}(r) + \big[\tilde{B}(r)^* P(r) \\
&\qquad + \tilde{S}_0(r)\big]^* \tilde{R}_{00}(r)^\dagger \tilde{S}_0(r) \Big\} \Phi(r,t)\xi' \Big\rangle dr,
\end{aligned}
$$

which yields

$$
\begin{aligned}
&\langle [\tilde{P}(t) - P(t)]\xi, \xi' \rangle \\
&= \int_t^T \langle \tilde{B}(r)\Theta^*(r)\Phi(r,t)\xi', [\tilde{P}(r) - P(r)]\Phi_\Theta(r,t)\xi \rangle dr, \\
&\quad \text{for any } \xi, \xi' \in M^2.
\end{aligned}
$$

Thus $P(t) = \tilde{P}(t)$ for all $t \in [s, T]$.

The uniqueness of solution to (3.3.22) can be guranteed by the equivalence among them. Finally we prove that (3.3.23)(a) is equivalent to (3.3.23)(b). Let $\eta(\cdot)$ and $\tilde{\eta}(\cdot)$ satisfy (3.3.23)(a) and (3.3.23)(b), respectively. Using (3.3.24), for any $\xi \in M^2$, we have

$$\begin{aligned}
\langle \eta(t), \xi \rangle &= \Big\langle \tilde{g}, \Phi(T,t)\xi + \int_t^T \Phi_\Theta(T,r)\tilde{B}(r)\Theta^*(r)\Phi(r,t)\xi dr \Big\rangle \\
&\quad + \int_t^T \Big\langle \tilde{q}(r) + P(r)\tilde{b}(r) + \Theta^*(r)^*\tilde{\rho}_0(r), \Phi(r,t)\xi \\
&\quad + \int_t^r \Phi_\Theta(r,\alpha)\tilde{B}(\alpha)\Theta^*(\alpha)\Phi(\alpha,t)\xi d\alpha \Big\rangle dr \\
&= \langle \tilde{g}, \Phi(T,t)\xi \rangle + \int_t^T \langle \Phi(r,t)\xi, P(r)\tilde{b}(r) + \tilde{q}(r) + \Theta^*(r)^*\tilde{\rho}_0(r) \rangle dr \\
&\quad + \int_t^T \langle \eta(r), \tilde{B}(r)\Theta^*(r)\Phi(r,t)\xi \rangle dr.
\end{aligned}$$

Hence we deduce

$$\langle \eta(t) - \tilde{\eta}(t), \xi \rangle = \int_t^T \langle \eta(r) - \tilde{\eta}(r), \tilde{B}(r)\Theta^*(r)\Phi(r,t)\xi \rangle dr,$$

which implies $\eta(t) = \tilde{\eta}(t)$. This completes the proof of Lemma 3.3.8.

□

Now we give the sufficient and necessary conditions of the closed-loop solvability for Problem (P-NCD).

Theorem 3.3.9 *Let* (H3.3) *hold.* $B_0(\cdot)$, $R_{00}(\cdot)$, $R_{00}^\dagger(\cdot)$, $S_{00}(\cdot)$, $S_{01}(\cdot,\cdot)$ *are continuous,* $\tilde{S}_0^*[I - \tilde{R}_{00}^\dagger \tilde{R}_{00}] = 0$, *then* $(\Theta^*(\cdot), v^*(\cdot))$ *is the optimal closed-loop strategy of* Problem (P-NCD) *on* $[s,T]$ *if and only if*

(i) $(\Theta^*(\cdot), v^*(\cdot))$ *is given by (3.3.25), where for any* $\xi \in M^2$, $P(\cdot)$ *satisfies the integral Riccati equation (3.3.22) and* $\eta(\cdot)$ *satisfies the integral Eq. (3.3.23),*
(ii) $\tilde{R}_{00} \geqslant 0$, $\mathcal{R}(\tilde{B}^*P + \tilde{S}_0) \subseteq \mathcal{R}(\tilde{R}_{00})$, $\mathcal{R}(\tilde{B}^*\eta + \tilde{\rho}_0) \subseteq \mathcal{R}(\tilde{R}_{00})$,
(iii) $(\Theta^*(\cdot), v^*(\cdot)) \in \mathcal{Q}[s,T]$.

In the case the value function is

$$\begin{aligned}
V(s,\xi) = \int_s^T &\Big\{ 2\langle \eta(t), \tilde{b}(t) \rangle_{M^2} - \langle \tilde{R}_{00}(t)^\dagger [\tilde{B}(t)^*\eta(t) + \tilde{\rho}_0(t)], \tilde{B}(t)^*\eta(t) \\
&+ \tilde{\rho}_0(t) \rangle \Big\} dt + \langle P(s)\xi, \xi \rangle_{M^2} + 2\langle \eta(s), \xi \rangle_{M^2}.
\end{aligned}$$

Proof If $\Theta^*(\cdot)$ is given by (3.3.25)(a), then from $\mathcal{R}(\tilde{B}^*P + \tilde{S}_0) \subseteq \mathcal{R}(\tilde{R}_{00})$, Proposition 2.2.3 and Lemma 3.3.8, P is strongly continuous, thus $\Theta^*(\cdot) \in \Xi(s,T)$. Similarly from (3.3.23)(b) we have $\eta(\cdot) \in C([s,T]; M^2)$, thus $\tilde{R}_{00}^\dagger(\tilde{B}^*\eta + \tilde{\rho}_0) \in$

$L^2(s, T; \mathbb{R}^m)$, which implies $v^*(\cdot) \in L^2(s, T; \mathbb{R}^m)$. The sufficiency can be proved using Theorem 3.3.7, thus we only need to prove the necessity. As Problem (P-NCD) is closed-loop solvable on $[s, T]$, Problem (P-NCD) is closed-loop solvable on $[r, T]$ for any $r \in [s, T)$, by the proof in Theorem 3.3.5 we have

$$\begin{aligned}\int_r^T \Big\{&\langle \tilde{R}_{00}(t)v(t), v(t)\rangle + 2\langle [\tilde{R}_{00}(t)\Theta^*(t) + \tilde{B}(t)^* P(t) \\ &+\tilde{S}_0(t)]z(t), v(t)\rangle\Big\}dt \geqslant 0, \quad \forall\, v(\cdot) \in L^2(r, T; \mathbb{R}^m),\ \xi \in M^2,\end{aligned} \tag{3.3.55}$$

where

$$z(t) = \Phi_\Theta(t, r)\xi + \int_r^t \Phi_\Theta(t, \beta)\tilde{B}(\beta)v(\beta)d\beta, \quad t \in [r, T].$$

Choose $\xi \in M^2$ and $v_k(t) = \frac{1}{k}v(t)$, $v(\cdot) \in L^2(s, T; \mathbb{R}^m)$, then

$$z_k(t) = \Phi_\Theta(t, r)\xi + \frac{1}{k}\int_r^t \Phi_\Theta(t, \beta)\tilde{B}(\beta)v(\beta)d\beta, \quad t \in [r, T].$$

Let $\xi_0(t) := \Phi_\Theta(t, r)\xi$. Apparently

$$\sup_{r\leqslant t\leqslant T} \|z_k(t) - \xi_0(t)\|_{M^2} \to 0, \quad \text{as } k \to \infty. \tag{3.3.56}$$

From (3.3.55) we have

$$\begin{aligned}\overline{\lim_{k\to\infty}}\, k \int_r^T \Big\{&\langle \tilde{R}_{00}(t)v_k(t), v_k(t)\rangle + 2\langle [\tilde{R}_{00}(t)\Theta^*(t) + \tilde{B}(t)^* P(t) \\ &+\tilde{S}_0(t)]z_k(t), v_k(t)\rangle\Big\}dt \geqslant 0,\ \forall\, v(\cdot) \in L^2(s, T; \mathbb{R}^m),\ \xi \in M^2.\end{aligned}$$

It follows that

$$\begin{aligned}\overline{\lim_{k\to\infty}} \int_r^T \langle [\tilde{R}_{00}(t)\Theta^*(t) + \tilde{B}(t)^* P(t) + \tilde{S}_0(t)]z_k(t), v(t)\rangle dt \geqslant 0, \\ \forall\, v(\cdot) \in L^2(s, T; \mathbb{R}^m),\ \xi \in M^2,\end{aligned}$$

which and (3.3.56) imply that

$$\begin{aligned}\int_r^T \langle [\tilde{R}_{00}(t)\Theta^*(t) + \tilde{B}(t)^* P(t) + \tilde{S}_0(t)]\xi_0(t), v(t)\rangle dt \geqslant 0, \\ \forall\, v(\cdot) \in L^2(s, T; \mathbb{R}^m),\ \xi \in M^2.\end{aligned}$$

Choose $v(t) = v\mathbf{1}_{[r,r+\varepsilon]}(t)$, $v \in \mathbb{R}^m$, then

$$\int_r^{r+\varepsilon} \langle [\tilde{R}_{00}(t)\Theta^*(t) + \tilde{B}(t)^* P(t) + \tilde{S}_0(t)]\xi_0(t), v\rangle dt \geqslant 0, \quad \forall\, v \in \mathbb{R}^m,\ \xi \in M^2.$$

Noting

$$\begin{aligned}
&\lim_{\varepsilon\to 0} \frac{1}{\varepsilon}\left| \int_r^{r+\varepsilon} \Big\langle \big[\tilde{R}_{00}(t)\Theta^*(t) + \tilde{B}(t)^* P(t) + \tilde{S}_0(t)\big](\xi_0(t) - \xi), v\Big\rangle dt \right| \\
&\leqslant K \lim_{\varepsilon\to 0} \sup_{r\leqslant t\leqslant r+\varepsilon} \|\Phi_\Theta(t,r)\xi - \xi\|_{M^2} = 0,
\end{aligned}$$

where K is a constant, thus we obtain

$$\lim_{\varepsilon\to 0} \frac{1}{\varepsilon} \int_r^{r+\varepsilon} \langle [\tilde{R}_{00}(t)\Theta^*(t) + \tilde{B}(t)^* P(t) + \tilde{S}_0(t)]\xi, v\rangle dt = 0, \quad \forall\, v \in \mathbb{R}^m,\ \xi \in M^2.$$

By the arbitrariness of ξ and v, it follows that

$$\tilde{R}_{00}(\tau)\Theta^*(\tau) + \tilde{B}(\tau)^* P(\tau) + \tilde{S}_0(\tau) = 0, \quad \text{a.e. } \tau \in [s,T), \tag{3.3.57}$$

which implies that (3.3.25)(a) holds and

$$\mathcal{R}(\tilde{B}^* P + \tilde{S}_0) \subseteq \mathcal{R}(\tilde{R}_{00}), \quad \text{a.e. } \tau \in [s,T).$$

Finally by (3.3.57) and Theorem 3.3.6, we have

$$\tilde{R}_{00}(\tau)v^*(\tau) + \tilde{B}(\tau)^*\eta(\tau) + \tilde{\rho}_0(\tau) = 0, \quad \text{a.e. } \tau \in [s,T].$$

This completes the proof of Theorem 3.3.9. □

Combining the above theorem and Theorem 3.2.9, we find the relationship between the closed-loop solvability and the existence of the closed-loop representation of open-loop optimal control for Problem (P-NCD).

Corollary 3.3.10 *Let* (H3.3) *hold. Suppose* $R_{00} > 0$, $\tilde{Q} \geqslant 0$, $\tilde{G} \geqslant 0$, $S_{00}, S_{01} = 0$, *then* $(\Theta^*(\cdot), v^*(\cdot))$ *is the optimal closed-loop strategy of* Problem (P_0-NCD) *on* $[s,T]$ *if and only if* $u^*(\cdot) = \Theta^*(\cdot)\mathbf{X}^*(\cdot) + v^*(\cdot)$ *is the closed-loop representation of open-loop optimal control. In this case,*

$$\Theta^*(t) = -\tilde{R}_{00}(t)^{-1}\tilde{B}(t)^* P(t), \quad v^*(t) = 0, \quad \text{a.e. } t \in [s,T],$$

where for any $\xi \in M^2$, $P(\cdot)$ satisfies the following integral Riccati equations:

$$
\begin{aligned}
(a)\quad P(t)\xi &= \Phi(T,t)^*\tilde{G}\Phi(T,t)\xi + \int_t^T \Phi(r,t)^*\Big\{\tilde{Q}(r) - \big[\tilde{B}(r)^*P(r)\big]^* \\
&\quad \times \tilde{R}_{00}(r)^{-1}\big[\tilde{B}(r)^*P(r)\big]\Big\}\Phi(r,t)\xi dr, \quad t \in [s,T],\\
(b)\quad P(t)\xi &= \Phi(T,t)^*\tilde{G}\Phi_\Theta(T,t)\xi + \int_t^T \Phi(r,t)^*\tilde{Q}(r)\Phi_\Theta(r,t)\xi dr, \ t \in [s,T],\\
(c)\quad P(t)\xi &= \Phi_\Theta(T,t)^*\tilde{G}\Phi_\Theta(T,t)\xi + \int_t^T \Phi_\Theta(r,t)^*\Big\{\tilde{Q}(r) + \big[\tilde{B}(r)^*P(r)\big]^* \\
&\quad \times \tilde{R}_{00}(r)^{-1}\big[\tilde{B}(r)^*P(r)\big]\Big\}\Phi_\Theta(r,t)\xi dr, \quad t \in [s,T],
\end{aligned}
$$

with $\Phi_\Theta(\cdot,\cdot)$ being the solution to the following integral equation:

$$
\Phi_\Theta(t,s)\xi = \Phi(t,s)\xi - \int_s^t \Phi(t,r)\tilde{B}(r)\tilde{R}_{00}(r)^{-1}\tilde{B}(r)^*P(r)\Phi_\Theta(r,s)\xi dr.
$$

And the value function is

$$
V(s,\xi) = \big\langle P(s)\xi, \xi\big\rangle_{M^2}.
$$

In the following example, the LQ problem admits an open-loop optimal control, but is not closed-loop solvable. Consider the following ODDE:

$$
\begin{cases} \dot{X}(t) = X(t-\delta) + u(t), & \text{a.e. } t \in [0,T],\\ X(t) = 0, & t \in [-\delta, 0], \end{cases} \tag{3.3.58}
$$

and the cost functional:

$$
J(u(\cdot)) = \int_0^T |X(t)|^2 dt.
$$

Noting that $\tilde{B}(r)u : \mathbb{R}^m \to M^2$, $u \mapsto (u^\top, 0)^\top$, $\tilde{R}_{00} = 0$, we have $\mathcal{R}(\tilde{B}^*P) \subsetneq \mathcal{R}(\tilde{R}_{00})$, this is a violation of Theorem 3.3.9 (ii) and thus the problem is not closed-loop solvable. However, $u^*(\cdot) = 0$ is the open-loop optimal control. In fact, $X^*(\cdot) = 0$ is the solution to (3.3.58). By (3.1.28), $p_1^0(\cdot) = 0$, hence the stationarity condition (3.1.29) holds. It is apparent to verify that (3.1.30) also holds, thus Corollary 3.1.7 implies that $u^*(\cdot) = 0$ is the open-loop optimal control.

Finally we give a more straight expression of the closed-loop solvability for Problem (P-NCD). Consider the special state Eq. (3.1.13) with B_i, $B^0 = 0$, $i = 1, \cdots, N$, inspired by the similar heuristic derivation before Theorem 3.2.10, we introduce the following CMREs:

$$
\begin{cases}
\frac{d}{dt}\mathscr{E}^0(t) + A_0(t)^\top \mathscr{E}^0(t) + \mathscr{E}_0(t)A_0(t) + Q_{00}(t) + \mathscr{E}^1(t,0) + \mathscr{E}^1(t,0)^\top \\
\quad -\big[B_0(t)^\top \mathscr{E}^0(t)\big]^\top R_{00}(t)^\dagger \big[B_0(t)^\top \mathscr{E}^0(t)\big] - S_{00}(t)^\top R_{00}(t)^\dagger B_0(t)^\top \mathscr{E}^0(t) \\
\quad -\mathscr{E}^0(t)B_0(t)R_{00}(t)^\dagger S_{00}(t) - S_{00}(t)^\top R_{00}(t)^\dagger S_{00}(t) = 0, \ \text{a.e. } t, \\
\Big(\frac{\partial}{\partial t} - \frac{\partial}{\partial \theta}\Big)\mathscr{E}^1(t,\theta) + A_0(t)^\top \mathscr{E}^1(t,\theta) + \mathscr{E}^0(t)\Big[\sum_{i=1}^{N-1} A_i(t)\hat{\delta}(\theta-\theta_i) \\
\quad +A^0(t,\theta)\Big] + \mathscr{E}^2(t,0,\theta) + Q_{10}(t,\theta)^\top - \mathscr{E}^0(t)B_0(t)R_{00}(t)^\dagger B_0(t)^\top \mathscr{E}^1(t,\theta) \\
\quad -S_{00}(t)^\top R_{00}(t)^\dagger B_0(t)^\top \mathscr{E}^1(t,\theta) - \mathscr{E}^0(t)B_0(t)R_{00}(t)^\dagger S_{01}(t,\theta) \\
\quad -S_{00}(t)^\top R_{00}(t)^\dagger S_{01}(t,\theta) = 0, \quad \text{a.e. } t,\theta, \\
\Big(\frac{\partial}{\partial t} - \frac{\partial}{\partial \theta} - \frac{\partial}{\partial \alpha}\Big)\mathscr{E}^2(t,\theta,\alpha) + A^0(t,\theta)^\top \mathscr{E}^1(t,\alpha) + \mathscr{E}^1(t,\theta)^\top A^0(t,\alpha) \\
\quad +Q_{11}(t,\alpha,\theta) - \big[B_0(t)^\top \mathscr{E}^1(t,\theta)\big]^\top R_{00}(t)^\dagger \big[B_0(t)^\top \mathscr{E}^1(t,\alpha)\big] \\
\quad +\sum_{i=1}^{N-1}\Big[A_i(t)^\top \mathscr{E}^1(t,\alpha)\hat{\delta}(\theta-\theta_i) + \mathscr{E}^1(t,\theta)^\top A_i(t)\hat{\delta}(\alpha-\theta_i)\Big] \\
\quad -S_{01}(t,\theta)^\top R_{00}(t)^\dagger B_0(t)^\top \mathscr{E}^1(t,\alpha) - \mathscr{E}^1(t,\theta)^\top B_0(t)R_{00}(t)^\dagger S_{01}(t,\alpha) \\
\quad -S_{01}(t,\theta)^\top R_{00}(t)^\dagger S_{01}(t,\alpha) = 0, \quad \text{a.e. } t,\theta,\alpha, \\
\mathscr{E}^0(T) = G_{00}, \ \mathscr{E}^1(t,-\delta) = \mathscr{E}^0(t)A_N(t), \ \mathscr{E}^1(T,\theta) = G_{10}(\theta)^\top, \\
\mathscr{E}^2(t,\theta,-\delta) = \mathscr{E}^1(t,\theta)^\top A_N(t), \ \mathscr{E}^2(t,-\delta,\alpha) = A_N(t)^\top \mathscr{E}^1(t,\alpha), \\
\mathscr{E}^2(T,\theta,\alpha) = G_{11}(\alpha,\theta),
\end{cases}
\tag{3.3.59}
$$

and the following CPDEs:

$$
\begin{cases}
\frac{d}{dt}\eta^0(t) + A_0(t)^\top \eta^0(t) - \mathscr{E}^0(t)B_0(t)R_{00}(t)^\dagger B_0(t)^\top \eta^0(t) \\
\quad -S_{00}(t)^\top R_{00}(t)^\dagger B_0(t)^\top \eta^0(t) + \eta^1(t,0) + q_0(t) + \mathscr{E}^0(t)b(t) \\
\quad -\mathscr{E}^0(t)B_0(t)R_{00}(t)^\dagger \rho_0(t) - S_{00}(t)^\top R_{00}(t)^\dagger \rho_0(t) = 0, \ \text{a.e. } t, \\
\Big(\frac{\partial}{\partial t} - \frac{\partial}{\partial \theta}\Big)\eta^1(t,\theta) + \Big[\sum_{i=1}^{N-1} A_i(t)^\top \hat{\delta}(\theta-\theta_i) + A^0(t,\theta)^\top\Big]\eta^0(t) \\
\quad -\mathscr{E}^1(t,\theta)^\top B_0(t)R_{00}(t)^\dagger B_0(t)^\top \eta^0(t) - S_{01}(t,\theta)^\top R_{00}(t)^\dagger B_0(t)^\top \eta^0(t) \\
\quad +q_1(t,\theta) + \mathscr{E}^1(t,\theta)^\top b(t) - \mathscr{E}^1(t,\theta)^\top B_0(t)R_{00}(t)^\dagger \rho_0(t) \\
\quad -S_{01}(t,\theta)^\top R_{00}(t)^\dagger \rho_0(t) = 0, \quad \text{a.e. } t,\theta, \\
\eta^0(T) = g_0, \quad \eta^1(T,\theta) = g_1(\theta), \quad \eta^1(t,-\delta) = A_N(t)^\top \eta^0(t).
\end{cases}
\tag{3.3.60}
$$

Theorem 3.3.11 *Let* (H3.3)(ii) *hold. Assume* $A_i(\cdot) \in L^\infty(0, T+\delta; \mathbb{R}^{n\times n})$, $A^0(\cdot,\cdot) \in L^\infty([0, T+\delta]\times[-\delta, 0]; \mathbb{R}^{n\times n})$, $b(\cdot) \in L^2(0, T; \mathbb{R}^n)$, $R_{00} \geqslant 0$, $B_0(\cdot)$ *and* $R_{00}(\cdot)^\dagger$ *are continuous. Let* $\mathscr{E}^0(t)$, $\mathscr{E}^1(t,\theta)$, $\mathscr{E}^2(t,\theta,\alpha)$, $\eta^0(t)$, $\eta^1(t,\theta)$, $t \in [s, T]$, $\theta, \alpha \in [-\delta, 0]$, *be absolutely continuous satisfying the Eqs.* (3.3.59)–(3.3.60), *and* $\mathscr{E}^0(t) = \mathscr{E}^0(t)^\top$, $\mathscr{E}^2(t,\theta,\alpha) = \mathscr{E}^2(t,\alpha,\theta)^\top$, *and*

$$\begin{cases} B_0(t)^\top \mathscr{E}^0(t)x + B_0(t)^\top \int_{-\delta}^0 \mathscr{E}^1(t,\theta)\varphi(\theta)d\theta + S_{00}(t)x \\ \quad + \int_{-\delta}^0 S_{01}(t,\theta)\varphi(\theta)d\theta \in \mathcal{R}(R_{00}(t)), \ \forall\, x \in \mathbb{R}^n, \ \varphi \in L^2(-\delta, 0; \mathbb{R}^n), \\ B_0(t)^\top \eta^0(t) + \rho_0(t) \in \mathcal{R}(R_{00}(t)). \end{cases} \tag{3.3.61}$$

Let $(\Theta^*(\cdot), v^*(\cdot))$ *be given by*

$$\begin{cases} \Theta^*(t)\xi = -R_{00}(t)^\dagger\Big[B_0(t)^\top \mathscr{E}^0(t)x + B_0(t)^\top \int_{-\delta}^0 \mathscr{E}^1(t,\theta)\varphi(\theta)d\theta \\ \qquad + S_{00}(t)x + \int_{-\delta}^0 S_{01}(t,\theta)\varphi(\theta)d\theta\Big] \\ \qquad + \big[I - R_{00}(t)^\dagger R_{00}(t)\big]\theta(t)\xi, \theta(\cdot) \in \Xi(s,T), \forall\, \xi = \begin{pmatrix} x \\ \varphi \end{pmatrix} \in M^2, \\ v^*(t) = -R_{00}(t)^\dagger\big[B_0(t)^\top \eta^0(t) + \rho_0(t)\big] + \big[I - R_{00}(t)^\dagger R_{00}(t)\big]\varsigma(t), \\ \qquad\qquad \varsigma(\cdot) \in L^2(s, T; \mathbb{R}^m). \end{cases} \tag{3.3.62}$$

Then $(\Theta^*(\cdot), v^*(\cdot))$ *is the optimal closed-loop strategy for* Problem (P-NCD) *with the following state equation:*

$$\begin{cases} \dot X(t) = \sum_{i=0}^N A_i(t)X(t+\theta_i) + \int_{-\delta}^0 A^0(t,\theta)X(t+\theta)d\theta \\ \qquad + B_0(t)u(t) + b(t), \ \text{a.e. } t \in [s, T], \\ X(s) = x, \quad X(t) = \varphi(t-s), \quad t \in [s-\delta, s), \end{cases} \tag{3.3.63}$$

and the value function is

$$\begin{aligned} V(s,x,\varphi(\cdot)) &= \big\langle \mathscr{E}^0(s)x + 2\eta^0(s), x\big\rangle + 2\int_{-\delta}^0 \big\langle \mathscr{E}^1(s,\theta)\varphi(\theta), x\big\rangle d\theta \\ &\quad + \int_{-\delta}^0 \int_{-\delta}^0 \big\langle \mathscr{E}^2(s,\theta,\alpha)\varphi(\alpha), \varphi(\theta)\big\rangle d\alpha d\theta \\ &\quad + 2\int_{-\delta}^0 \big\langle \eta^1(s,\theta), \varphi(\theta)\big\rangle d\theta + \int_s^T \Big\{2\big\langle \eta^0(t), b(t)\big\rangle \\ &\quad - \big\langle R_{00}(t)^\dagger\big[B_0(t)^\top \eta^0(t) + \rho_0(t)\big], B_0(t)^\top \eta^0(t) + \rho_0(t)\big\rangle\Big\} dt. \end{aligned} \tag{3.3.64}$$

Proof For any admissible control $u(\cdot) \in L^2(s, T; \mathbb{R}^m)$, let $X(\cdot)$ be the corresponding state satisfying (3.3.63), define

$$\begin{aligned}\Gamma(t) := &\big\langle \mathscr{E}^0(t)X(t) + 2\eta^0(t), X(t)\big\rangle + 2\int_{-\delta}^{0} \big\langle \mathscr{E}^1(t,\theta)X(t+\theta), X(t)\big\rangle d\theta \\ &+ \int_{-\delta}^{0}\int_{-\delta}^{0} \big\langle \mathscr{E}^2(t,\theta,\alpha)X(t+\alpha), X(t+\theta)\big\rangle d\alpha d\theta \\ &+2\int_{-\delta}^{0} \big\langle \eta^1(t,\theta), X(t+\theta)\big\rangle d\theta + 2\int_{t}^{T} \Big\{2\big\langle \eta^0(r), b(r)\big\rangle \\ &-\big\langle R_{00}(r)^{\dagger}\big[B_0(r)^{\top}\eta^0(r) + \rho_0(r)\big], B_0(r)^{\top}\eta^0(r) + \rho_0(r)\big\rangle\Big\} dr.\end{aligned}$$

By (3.3.59)–(3.3.62), with some computations we derive

$$\begin{aligned}&\frac{d}{dt}\Gamma(t) + \big\langle Q_{00}(t)X(t), X(t)\big\rangle + 2\int_{-\delta}^{0} \big\langle Q_{10}(t,\theta)^{\top}X(t+\theta), X(t)\big\rangle d\theta \\ &+ \int_{-\delta}^{0}\int_{-\delta}^{0} \big\langle Q_{11}(t,\theta,\theta')X(t+\theta), X(t+\theta')\big\rangle d\theta' d\theta + 2\big\langle S_{00}(t)X(t), u(t)\big\rangle \\ &+2\int_{-\delta}^{0} \big\langle S_{01}(t,\theta)X(t+\theta), u(t)\big\rangle d\theta + \big\langle R_{00}(t)u(t), u(t)\big\rangle + 2\big\langle q_0(t), X(t)\big\rangle \\ &+2\int_{-\delta}^{0} \big\langle q_1(t,\theta), X(t+\theta)\big\rangle d\theta + 2\big\langle \rho_0(t), u(t)\big\rangle \\ =&\Big\langle R_{00}(t)\Big[u(t) - \Theta^*(t)\begin{pmatrix} X(t) \\ X_t \end{pmatrix} - v^*(t)\Big], u(t) - \Theta^*(t)\begin{pmatrix} X(t) \\ X_t \end{pmatrix} - v^*(t)\Big\rangle, \\ &\qquad\qquad\qquad\qquad\qquad\qquad \text{a.e. } t \in [s, T].\end{aligned}$$

Similar to the proof of Theorem 3.2.10, the proof is completed. □

Remark 3.3.12 Assume that $A_i = 0$, $i = 1, \cdots, N-1$ and $G_{00}, G_{10}, G_{11} = 0$. Then, the CMRE (3.3.59) admits a unique solution. The detailed proof can be referred to Alekal et al. [1]. Hence, the CPDE (3.3.60) also admits a unique solution. Moreover, let $B_i = 0$, $i = 1, \cdots, N$, $R_{00} > 0$, $B^0, S_{00}, S_{01}, S_{10}, S_{11}, R_{10}, R_{11} = 0$, then (3.3.59) is consistent with the CMRE (3.2.36).

Remark 3.3.13 Consider Problem (P-NCD) without the delays, i.e. A_i, A^0, Q_{10}, Q_{11}, S_{01}, q_1, G_{10}, G_{11}, $g_1 = 0$, $i = 1, \cdots, N$, then Problem (P-NCD) becomes the optimal control problem in Sun-Yong [12] where the diffusion term is absent. In this case M^2 reduces to $\mathbb{R}^n$ and $\Theta^*(t)$ is no longer an operator, which reduces to an $\mathbb{R}^{m\times n}$-valued matrix. Moreover, (3.3.59)–(3.3.62) reduce to the Riccati (2.4.1) and the BODE (2.4.3) in [12], the above theorem becomes the sufficient part of Theorem 2.4.3 in it.

In the end of this section, ignoring the delay of the executive control, we give the closed-loop solvability of Problem (Ex). In this case, $k_4 = 0$. Introduce the following IOREs:

$$
\begin{aligned}
&(a)\ P(t)\xi = \int_t^T \Phi(r,t)^*\Big\{\tilde{Q}(r) - \big[\tilde{B}(r)^*P(r)\big]^*\tilde{R}_{00}(r)^\dagger\big[\tilde{B}(r)^*P(r)\big]\Big\} \\
&\qquad\qquad \times\Phi(r,t)\xi dr, \qquad t\in[s,T], \\
&(b)\ P(t)\xi = \int_t^T \Phi_\Theta(r,t)^*\big[\tilde{Q}(r) + \Theta^*(r)^*\tilde{R}_{00}(r)\Theta^*(r)\big]\Phi_\Theta(r,t)\xi dr, \\
&\qquad\qquad\qquad\qquad\qquad\qquad t\in[s,T], \\
&(c)\ P(t)\xi = \int_t^T \Phi(r,t)^*\tilde{Q}(r)\Phi_\Theta(r,t)\xi dr,\ t\in[s,T],
\end{aligned}
\tag{3.3.65}
$$

and the following BIEEs:

$$
\begin{aligned}
&(a)\quad \eta(t) = \int_t^T \Phi_\Theta(r,t)^*\big[\tilde{q}(r) + P(r)\tilde{b}(r)\big]dr, \quad t\in[s,T], \\
&(b)\quad \eta(t) = \int_t^T \Phi(r,t)^*\big[P(r)\tilde{b}(r) + \Theta^*(r)^*\tilde{B}(r)^*\eta(r) \\
&\qquad\qquad +\tilde{q}(r)\big]dr, \quad t\in[s,T],
\end{aligned}
\tag{3.3.66}
$$

where $\Phi(\cdot,\cdot)$ is the solution to (3.3.24). In the above equations,

$$
\begin{aligned}
&(a)\quad \Theta^*(t) = -\tilde{R}_{00}(t)^\dagger\tilde{B}(t)^*P(t) \\
&\qquad\qquad +\big[I - \tilde{R}_{00}(t)^\dagger\tilde{R}_{00}(t)\big]\theta(t), \quad \text{a.e. } t\in[s,T], \\
&(b)\quad v^*(t) = -\tilde{R}_{00}(t)^\dagger\tilde{B}(t)^*\eta(t) \\
&\qquad\qquad +\big[I - \tilde{R}_{00}(t)^\dagger\tilde{R}_{00}(t)\big]\varsigma(t), \quad \text{a.e. } t\in[s,T],
\end{aligned}
\tag{3.3.67}
$$

for any $\theta(\cdot)\in\Xi(s,T)$ and any $\varsigma(\cdot)\in L^2(s,T;\mathbb{R}^m)$.

Corollary 3.3.14 *Let $k_4 = 0$. Then $(\Theta^*(\cdot), v^*(\cdot))$ is the optimal closed-loop strategy of* Problem (Ex) *on $[s,T]$ if and only if*

(i) *$(\Theta^*(\cdot), v^*(\cdot))$ is given by (3.3.67), where for any $\xi\in M^2$, $P(\cdot)$ satisfies the integral Riccati equation (3.3.65) and $\eta(\cdot)$ satisfies the integral Eq. (3.3.66),*
(ii) *$r\geqslant 0$, $\mathcal{R}(\tilde{B}^*P)\subseteq\mathcal{R}(\tilde{R}_{00})$, $\mathcal{R}(\tilde{B}^*\eta)\subseteq\mathcal{R}(\tilde{R}_{00})$,*
(iii) *$(\Theta^*(\cdot), v^*(\cdot))\in\mathcal{Q}[s,T]$.*

In the case the value function is

$$
\begin{aligned}
V(s,\xi) = &\int_s^T \Big\{2\langle\eta(t),\tilde{b}(t)\rangle_{M^2} - \langle\tilde{R}_{00}(t)^\dagger\big[\tilde{B}(t)^*\eta(t)\big],\tilde{B}(t)^*\eta(t)\rangle\Big\}dt \\
&+\langle P(s)\xi,\xi\rangle_{M^2} + 2\langle\eta(s),\xi\rangle_{M^2}.
\end{aligned}
$$

Introduce the following CMREs:

$$
\begin{cases}
\frac{d}{dt}\mathscr{E}^0(t) - 2k_1\mathscr{E}^0(t) + q + 2\mathscr{E}^1(t,0) - r^\dagger k_3^2\mathscr{E}^0(t)^2 = 0, \text{ a.e. } t, \\
\Big(\frac{\partial}{\partial t} - \frac{\partial}{\partial \theta}\Big)\mathscr{E}^1(t,\theta) - k_1\mathscr{E}^1(t,\theta) + \mathscr{E}^2(t,0,\theta) \\
\quad -r^\dagger k_3^2\mathscr{E}^0(t)\mathscr{E}^1(t,\theta) = 0, \quad \text{a.e. } t,\theta, \\
\Big(\frac{\partial}{\partial t} - \frac{\partial}{\partial \theta} - \frac{\partial}{\partial \alpha}\Big)\mathscr{E}^2(t,\theta,\alpha) - r^\dagger k_3^2\mathscr{E}^1(t,\theta)\mathscr{E}^1(t,\alpha) = 0, \quad \text{a.e. } t,\theta,\alpha, \\
\mathscr{E}^0(T) = 0, \ \mathscr{E}^1(t,-\delta) = k_2\mathscr{E}^0(t), \ \mathscr{E}^1(T,\theta) = 0, \\
\mathscr{E}^2(t,\theta,-\delta) = k_2\mathscr{E}^1(t,\theta), \ \mathscr{E}^2(t,-\delta,\alpha) = k_2\mathscr{E}^1(t,\alpha), \\
\mathscr{E}^2(T,\theta,\alpha) = 0,
\end{cases}
\tag{3.3.68}
$$

and the following CPDEs:

$$
\begin{cases}
\frac{d}{dt}\eta^0(t) + k_1\eta^0(t) - r^\dagger k_3^2\mathscr{E}^0(t)\eta^0(t) \\
\quad +\eta^1(t,0) - aq + k_5\mathscr{E}^0(t)f(t-\delta) = 0, \text{ a.e. } t, \\
\Big(\frac{\partial}{\partial t} - \frac{\partial}{\partial \theta}\Big)\eta^1(t,\theta) - r^\dagger k_3^2\mathscr{E}^1(t,\theta)\eta^0(t) \\
\quad +k_5\mathscr{E}^1(t,\theta)f(t-\delta) = 0, \quad \text{a.e. } t,\theta, \\
\eta^0(T) = 0, \quad \eta^1(T,\theta) = 0, \quad \eta^1(t,-\delta) = k_2\eta^0(t).
\end{cases}
\tag{3.3.69}
$$

Corollary 3.3.15 *Assume $r \geqslant 0$. Let $\mathscr{E}^0(t)$, $\mathscr{E}^1(t,\theta)$, $\mathscr{E}^2(t,\theta,\alpha)$, $\eta^0(t)$, $\eta^1(t,\theta)$, $t \in [0,T]$, $\theta,\alpha \in [-\delta,0]$, be absolutely continuous satisfying the Eqs.* (3.3.68)–(3.3.69), *and*

$$
\begin{cases}
k_3\mathscr{E}^0(t)x + k_3\displaystyle\int_{-\delta}^0 \mathscr{E}^1(t,\theta)\varphi(\theta)d\theta \in \mathcal{R}(r), \ \forall\, x \in \mathbb{R}, \ \varphi \in L^2(-\delta,0;\mathbb{R}), \\
k_3\eta^0(t) \in \mathcal{R}(r).
\end{cases}
$$

Let $(\Theta^*(\cdot), v^*(\cdot))$ be given by

$$
\begin{cases}
\Theta^*(t)\xi = -r^\dagger\Big[k_3\mathscr{E}^0(t)x + k_3\displaystyle\int_{-\delta}^0 \mathscr{E}^1(t,\theta)\varphi(\theta)d\theta\Big] \\
\qquad\qquad +\big[1 - r^\dagger r\big]\theta(t)\xi, \theta(\cdot) \in \Xi(s,T), \forall\, \xi = \begin{pmatrix} x \\ \varphi \end{pmatrix} \in M^2, \\
v^*(t) = -r^\dagger k_3\eta^0(t) + \big[1 - r^\dagger r\big]\varsigma(t), \qquad \varsigma(\cdot) \in L^2(s,T;\mathbb{R}).
\end{cases}
$$

Then $(\Theta^*(\cdot), v^*(\cdot))$ is the optimal closed-loop strategy for Problem (Ex) *and the value function is* (3.3.64).

3.4 Technical Remarks

As in Sect. 2.3, the new control operator **B** makes Problem (EP) non-standard. **B** maps $\mathbb{R}^m$ to $M^2 \times V^*$ out Z, thus we can not use the control theory in [2, 7]. The domain of **B** has nothing to do with the domain of the generator operator **A**, thus we can also not apply the theory in [5, 6] to deal with the difficulties caused by **B**. Therefore, in this chapter, we use some other methods different from the existing literature.

In Sect. 3.1, we construct the adjoint equation by virtue of the explicit definitions of **B** and other operators, which results in the novel form of the adjoint Eq. (3.1.2), especially the last integral evolution equation. This is a feature of the infinite dimensional LQ optimal control problem originating from a control problem with control delays, compared with the general infinite dimensional LQ optimal control problem (referred to Li–Yong [7] and Lions [8]). When going back to the original delayed control problem (P) and characterizing the open-loop solvability, Lemmas 3.1.4 and 3.1.5 play key roles in the equivalence of Problem (P) and Problem (EP), and their proofs are referred to Theorems 2.1 and 3.1 in [3], respectively. Similar to (3.1.2), the adjoint Eq. (3.1.23) is a CPDE with a novel form, while in the existing literature the adjoint equations are a kind of anticipated backward ordinary differential equations.

In Sect. 3.2, we apply some approximation arguments to deal with **B**. To this end, the convergence results of Lemmas 3.2.4 and 3.2.5 are very crucial and their detailed proofs are referred to Lemma 1.2 and Lemma 1.3 in Ichikawa [4], respectively.

To complete this technical remark section, as well as to conclude this book, let us make a short summary. In this book, we consider a general delayed LQ optimal control problem and study its open-loop and closed-loop solvability, which makes the theory of delayed optimal control problems more systematic and getting closer to complete. It turns out that the feedback of the optimal control consists of four terms: the one proportional to the current value of the state, the one not related to the state and the control which is used to handle the nonhomogeneous part in the state equation and the linear penalty in the cost functional, the other two involving the integral of the state and the control trajectories over the past time interval, respectively. The first two terms are similar to the classical feedback control, while the last two terms characterize the dependence on history. Due to the complex feedback structure, how to define the open-loop and the closed-loop solvability for Problem (P), is a crucial question of this book.

In Sects. 2.1 and 2.2, we lift the state equation of Problem (P) from $\mathbb{R}^n$ to an infinite dimensional Hilbert space, and transform the control problem (P) with delays into Problem (EP) without delays. Then we introduce three appropriate definitions: open-loop solvability (Definition 2.2.4), closed-loop solvability (Definition 2.2.8) and closed-loop representation of open-loop optimal control (Definition 2.2.9) for Problem (P).

We split three steps to characterize the open-loop and the closed-loop solvability:

- The first step is to obtain sufficient and necessary conditions of the open-loop solvability for Problem (P) (Theorems 3.1.1, 3.1.3 and 3.1.7), which is characterized by the solvability of the system of FBIEEs (3.1.2), the CPDEs (3.1.23) and the convexity condition (3.1.3). Because of the new control operators, the adjoint equations have very novel structure.
- The second step is to derive the closed-loop representation of open-loop optimal control for Problem (P_0) (Theorems 3.2.9 and 3.2.10), which is given through three equivalent IOREs (3.2.30) and the CMREs (3.2.36)–(3.2.38).
- The third step is to get the closed-loop solvability for Problem (P-NCD) (Theorems 3.3.9 and 3.3.11), which is characterized by three equivalent IOREs (3.3.22) and two equivalent BIEEs (3.3.23), or the CMREs (3.3.59) and the CPDEs (3.3.60).

The above three steps clearly characterize the open-loop and the closed-loop solvability for Problem (P). The close-loop solvability implies the open-loop solvability, and under proper assumptions, Problem (P) is closed-loop solvable if and only if it admits the closed-loop representation of open-loop optimal control.

Acknowledgments This work was partially supported by National Key *R&D*. Program of China (2022YFA1006104), National Natural Science Foundations of China (12471419, 11831010, 12271304), Shandong Provincial Natural Science Foundations (ZR2022JQ01, ZR2024ZD35), and NSF Grants DMS-1812921 and DMS-2305475.

References

1. Alekal, Y., Brunovský, P., Chyung, D.H., Lee, L.E.: The quadratic problem for systems with time delay. IEEE Trans. Automat. Control **16**, 673–687 (1971)
2. Curtain, R.F., Pritchard, A.J.: Infinite Dimensional Linear Systems Theory. Lecture Notes in Control and Information Sciences, vol. 8. Springer-Verlag, Berlin-New York (1978)
3. Delfour, M.C.: State theory of linear hereditary differential systems. J. Math. Anal. Appl. **60**, 8–35 (1977)
4. Ichikawa, A.: Quadratic control of evolution equations with delays in control. SIAM J. Control Optim. **20**, 645–668 (1982)
5. Lasiecka, I., Triggiani, R.: Control Theory for Partial Differential Equations: Continuous and Approximation Theories I: Abstract Parabolic Systems. Cambridge University Press, Cambridge (2000)
6. Lasiecka, I., Triggiani, R.: Control Theory for Partial Differential Equations: Continuous and Approximation Theories II: Abstract Hyperbolic-Like Systems over a Finite Time Horizon. Cambridge University Press, Cambridge (2000)
7. Li, X., Yong, J.M.: Optimal Control Theory for Infinite Dimensional Systems. Birkhäuser, Boston (1995)
8. Lions, J.L.: Optimal Control of Systems Governed by Partial Differential Equations. Springer-Verlag, Berlin/Heidelberg/New York (1971). Translated by S. K. Mitter
9. Lü, Q.: Stochastic linear quadratic optimal control problems for mean-field stochastic evolution equations. ESAIM Control Optim. Calc. Var. **26**, 127 (2020)

10. Pontryagin, L.S., Boltyanskii, V.G., Gamkrelidze, R.V., Mishchenko, E.F.: The Mathematical Theory of Optimal Processes. Interscience, New York (1962)
11. Renardy, M., Rogers, R.C.: An Introduction to Partial Differential Equations. Springer, New York (2004)
12. Sun, J.R., Yong, J.M.: Stochastic Linear-Quadratic Optimal Control Theory: Open-Loop and Closed-Loop Solutions. Springer, Berlin (2020)
13. Yong, J.M., Zhou, X.Y.: Stochastic Controls: Hamiltonian Systems and HJB Equations. Springer-Verlag, New York (1999)

Index

© The Author(s), under exclusive license to Springer Nature Singapore Pte Ltd. 2025

W. Meng et al., *Time-Delayed Linear Quadratic Optimal Control Problems*, SpringerBriefs on PDEs and Data Science,
https://doi.org/10.1007/978-981-96-1897-2